THE BIG G UNIVERSE
and

Reverse-Vacuum Space Field Drive

e+h@g2.7c = FTL

PREFACE

Why does the apple fall to the ground Sir Isaac Newton the queen asked him?

Because - it has matured and is now too heavy for its stem to hold it anymore and there's nothing stopping it from between the tree branch and the ground.

It is *Rodney Kawecki's Gravity Theory* that will explain the mystery's left behind by physicists about the weight of things that have been placed on the back burner in the dictum of theoretical physics and explained to become confused with energy, mass and matter and may be the final explanation for a theory of gravity using the invisible forces that bind it. Thought to be a single particle or string partisan or membrane the gravity force is actually the activity cause and effects of forces that bind a planet and its atmosphere and divides it from the foreign regional astronomical celestials with outer space.

Rodney Kawecki, born in Los Angeles California studied advance modern physics for ten years researching his idea for a new modern theory about space. His idea was that faster than light space travel was possible *simply* based on weightlessness and for foremost characteristic's resistance rather than an attraction force. It became a lifelong adventure to discover the equations needed to improve his new ideas. That day came in 2007 in his first book on faster than light speed space technology where everything about faster than light space travel is exhibited for the purpose of technology and his thoughts that space is the 'beyonder foundation' and an important chapter in earth's future.

His work on modern physics opens the door to new possibilities in space aviation space flight and includes theories for faster than light space travel. This piece of literature is about gravity. An invisible force which on earth seems to retain a speed condition to any object free falling towards the planet's surface or in acceleration, *Rodney Kawecki* tries to open the door to why this phenomenon is a natural order of out planet's atmosphere. He defines the invisible forces that he believes act as elements for the gravity force which earlier scientist have over looked.

Table of Contents

Introduction

At a far distant edge of the great horizon a frantic little girl releases a little butterfly from out of the palms of her hands in the front of her. Catching sight of a gem of light she makes chase for it as it is said to travel through space at its own measurable time and distance. Afar lays the brightest star whose ray of light is only seeable because the universe rotates at its own orbiting angular singularity curvature through space. The stars dense material is extreme in the readings of the cosmos space element charts as it lays deepest in the space fabric surface grid. Even the densest material in the universe cannot fracture the infinite space grid it lays in nor pass through it to reach onto the other side of it. Trying to take short cuts in space is only the figment of the imagination.

Gravity decreases with altitude, since greater altitude means greater distance from the Earth's center. At the Interstellar stage it ceases to exist in the same form as a detectable prorogating wave.

Two stars of similar mass are in circular orbits about their center of mass. *(energy acting as a similar 'effect' and rolling due to their shapes)

Two stars of dissimilar mass are in circular orbits. Each will rotate about their common center of mass in a circle with the larger, the dominion mass having a smaller orbit. *(object masses are measured outside the sphere, whereas, ' gravitation ' is the actual effect of objects bulk energy mass, acting adjoins with its mass towards the smaller mass attracting it to a specific distance). Gravity is commonly confused with the magnetic force, which actually retains the balance of an objects position and not the distance between objects. The energy content in relativity assets to Newtonian attraction theory as well but two positive poles repel they don't attract either.

It can't be disputed that Gravity is in itself, a sort of planet element. But at this moment of time, science stops short of declaring gravity as only "on observation basis" because we could only see and describe its effects, but not its nature. Not yet, until now says Rodney Kawecki. Author of the first literature on faster than light space travel within the standard model of the universe and trait believed to be impossible by earlier physicists.

Newtonian mechanics assert that 9.8 meters per second is the interval allowed by the planet surface density for object free fall. No matter the mass weight of the item it will only accelerate at that interval no more no less.

This measured rate of increased velocity is the same to all object free fall.

Newtonian gravity takes this uniformity to explain gravity. It also further explains this uniformity due to attraction. This somewhat confuses some to think gravity is linked to magnetism hence energy mass but it's not.

Gravity actually works in the opposite. Rather than a force of attraction gravity repels similar matter a ten pound ball free falls from a plane. Along next to it a single pound ball. They both travel to the general surface at 9.8 m.p.s *(meters)*

Relativity asserts that the heavier ball has more energy there for the attraction should be more but it's not. The question then is why? The answer is that all surface objects that enter or are within the planet's atmosphere abide by that planet's atmosphere energy and atmosphere elements. A surface object haven laid in the planet's surface atmosphere grid retains the planets energy and element weight measurement. On a different planet that weight measurement may be different dependent to the sphere's density and grade elements. It is not all energy that defines the force of gravity on that planet.

The opposing argument Quanta Physics Theory raises with relativity and Newtonian physics about gravity is that gravity whether in space or within the planet's atmosphere area resist it does not attract.

The fact that celestials in space resist one another due to like energy resistance – allows the balance and foundation for the expanding universe space grid surface planets and galaxies have as they orbit space. The balance of mobility in space is enforced by the velocity of the universe in rotational orbit the spin and curvature that allows an advance force to mobilize galactic positioning in the fabric grid. The celestials do not fall in the grid – if they did they would pass through it and wormhole technology tells us in *Quanta Physics* that that is strictly almost impossible it there exist no evidence to the Rip Theory.

These are the renowned facts about our universe and its history in physics. Wormhole technology is premature data tech at this time and age trying to explain its potential. The possibility based in *quanta physics* for space travel short-cuts is the usage of cesium gaseous booster chemistry that has been known to advance object acceleration and velocity when an object passages through the applied cloud grid. Aside from cesium advance acceleration theory Rodney Kawecki only sides to "Instantaneous Faster than Light Quantum Travel Mechanics". The idea that he analyzed and researched that claims that wormhole short cutting and same like technology is not a valid theoretical space travel objective for future world space travel. The idea that he conformed to in his research about faster than light travel technology is that it can only be achieved by propulsion capacity in a free fall weightless space atmosphere dependent as well with the up and down positioning of the universe itself predictable navigated top to bottom free fall and acceleration assistive propulsion galactic highway passages abilities that might transpire in earths future. This is the basis for his writing in the theoretical space sciences reinventing space physics literature into a new and modern centurion age. Wormhole technology to materialize relativity would have to validate its claim for such a space travel technic by showing how an object can pass through the dark elemental' fabric grid surface that our planets, stars and galaxies assent upon. A test explaining the possibilities of passing through the dark elemental' fabric grid surface that our planets, stars and galaxies as physics shows these celestials ascend upon today only prove how they position in the gas cloud elemental grid not pass through it and that is a must. Could it be possible to achieve by the spacecraft's miniature celestial size that at a great enough velocity it asserts a small tiny wormhole through the grid and pass through it due to its arranged energy content bubble string appearance called a "wormhole" tunnel that inverts a small opening for a short while? If a galactic mass as great as a gigantic galaxy cannot tear or rip in surface of the space grid what can? It is at this point and time that special relativity and general relativity become a fable in science. (read: "Albert Einstein's Universe" by Rodney Kawecki 2011)

Newton asserts they are both the same but really shows us it's not.

Quanta Mechanics asserts that neither is the case. Planet Gravity acts as a weak yet equaling resistance within the planet's atmosphere. Facts show that gravity is designed by layers that measure at 9.8 m.p.s. each second not only by velocity but time interval. The object in free fall falls towards the planet surface not based on attraction or any magnetism of sort but due only to what the earth's natural atmosphere.

Some people have come to say through common knowledge about the theory of gravity addressing the saying that "There's no such thing as Gravity". The thing is that by investigation it' by well may be true. The fact is that the theory of gravity retains no real explanation. Energy and mass or energy mass retained in Albert Einstein's literature about gravity explains gravity with the idea that it's a bound force retaining a center of mass by which all things material abide. As energy mass type gravitational force his theory allows entering with the perception with electromagnetism theory that describes gravity with a balance using magnets which real doesn't work at all. Likewise as we review the foundation that governs energy or energy mass we discover that any center of gravity or center of mass the direction of activity centralizes towards the center of the object mass not outwards balancing a division between objects. It allows the increase of energy to localize an objects matter to increase and or decrease as relativity explains as infinite mass theory. It all detours any resolution about weight based on it as a physical theory.

Quanta Physics Theory as retained in its theology about space and space travel tries to explain gravity on the material scale aspect of the chemistry charts that measure the units of the elements as they exist at the molecular scale. What goes up must come down except in space of course. This new theory on gravity at the planetary scale abides that the weights and measures of a free falling object are physically based with rest mass at the earth's surface as the heaviest of an objects possible weight and as it approaches outer space at the gravity ring that governs the planet's atmosphere as the lightest possible weight of the object. As you will read later in this book how the free fall constant of an object falling towards the earth's surface increases in variables of 9.8 meters per second and the reason why it does this you will also discover a planet's atmosphere is layered by the planets rotation velocity combined with the planets invisible elements why this combination affects a falling object not seduced by energy but by the planets weight and motion.

Gravitation is most accurately described by the general theory of relativity by Einstein, in which the phenomenon itself is a consequence of the curvature of space. Weight of the sphere's by which space acts as a foundation that keeps the celestial bodies separately in motion.

This book you about to read are an explanation about what actually defines "The Gravity Force". Unlike latter definitions on gravity the chemistry elements do not define the actual characteristics that define falling objects or even for that matter celestial interstellar gravity. What I have designed on the subject of both planet gravitation and celestial gravity are unique and special as the boundaries explained about the gravity force are maintained solely on the balance of the cosmos as a whole.

What this means is that planet gravity and interstellar celestial gravity basically all entangle into a single activity which the causes and barriers existing as a universe all intertwine creating a single gravity mechanic. A good definition about this entanglement of the forces that recons elate the gravity all stem from a single activity.

In Star wars the sci-fi author entitled the inhabitants of this same entanglement calling it "The Force". Isaac Newton researched earth gravity as objects attraction to the planets material core he called the center of gravity. In 1904 Albert Einstein modernized Newton's attraction analogy with the internal forces of molecular energy that exist in all material matter. He theorized that the attraction of material surface objects acted as a weak force acting relative to the greater dominion of the larger celestial mass.

Quanta Physics Theory has researched the extreme idea about the universe and the invisible and noticeable forces that exist and balance the cosmos as a divine single unit. Rodney Kawecki goes further to try and discloses where, how and to what nature the material concept of reality existing through the presents of matter coming to arrive a material planetary universe divide. As you read this well natured book about gravity and nature of invisibility Kawecki opens a cosmic doorway to the unknown regions of the cosmos in parallel to the reality that matter, life and life essential elements arrive from a hidden region we call the fifth element called dark empty space. But the question is whether or not what we perceive as being empty is at all empty at all. Could the dark element from which planets, star and galaxy's reside as celestial sphere's and galaxy's and reform what we perceive as empty space now maintain a position and surface reside on a fabric we thought as empty space empty at all.

Einstein explained how celestials impress dark empty space and form surface grid characteristic and as such the celestial bodies reside positioned on a universal angular momentum velocity rotational track which angles the flat surface grid amongst which the planetary bodies galactic alike embed themselves bending, stretching and curving the space we deemed once as empty space. Called a universe pod – it rotates from a velocity much greater than light speed but retains a constant speed limit or close to consistent speed limit of all the galactic matter existing in it. Forming from a gigantic explosion weight impressed by the velocity rate of the bang that it started from and not able to rip, cut, or breaks through the fabric surface the matter curved in on itself. As an empty vacuum matter from the cosmic bang pushed at the extreme speed of the eruption that can be said to measure relevant to a growing embryus primitive egg event to overcome the fabric it manifested from and through its infinite grid shielding flexibility.

So the question is asked again where did everything making up the cosmos come from? Reviewing a balance of extreme invisible forces manifested by the pressure of a gigantic bang in a weightless terrain of empty space – it is these same forces that balance the orbit and rotation of the planetary bodies it manifested in the beginning. The universe maintains a specific and consistent rotation in space its outer rim or edges impress an elliptical path from which the galactic matter curved in on itself inside this gigantic galactic body's smaller celestial spheres orbit around great sunken suns embedded in the more dense fabric inside the heavy weighted galaxies. A celestial sphere's weight is measurable by the density the celestial impresses against the fabric in parallel to its rotation and orbitration inside the galaxy but only within the universe cosmos field terrain. Their biological clock orbits and rotations around these gigantic suns are counter played from colliding into the dominion sun impression in the fabric and smaller planets caught in the dominion stars deep impression of the fabric are counter played by the extreme forces of the universe's rotation allowing the smaller celestials from being pulled into the larger sun deities. This aligns the inner star systems and solar systems to survive and orbit the existing sun and rotate as the universe spin.

It is not dark space that the impression of celestial matter engraves a flat disk space grid but more so it is cosmos matter that empty space by the depth of its flexible fabrication it is dragged by the rotation of the acting celestial bodies in a system of bodies impressing a space curvature in the space field itself that its deep impression creates a flat orbiting grid plate. This has been shown in satellite journey observations that have traveled out pass the solar systems outer edge. To answer the question about gravity we have to look at the force continuum that exist through the universe. That at the planetary scale like earth gravity is an action of free fall coordinated by the planets spin at a velocity dragging the planet's atmosphere a long a very fast rotation and comparing that with the atmosphere element s raising and being pushed into layers that increase the subtle existence from which a planet's atmosphere air, heat become heavy invisible elements. As an object free falls towards the planet's surface it travels through potent layers that only allow passage to the surface at a specific speed.

Reviewing the facts that follow specifics about the space fabric dark element empty space and its relationship with an evolving and orbiting material deity we deem 'THE UNIVERSE' it's mechanics disclose evidence that energy or the density of matter spheres lay within that solidity whereas 'gravity' 'Celestial Gravity' all parts of the word is the push and play activity of orbiting, cycle continuality and rotations that organize a unique perpetual motion design forming weights and measures, balance and in some cases life deities in the cosmos. How mass energy relative to what earlier physicists believed to be the source chemistry of gravitation actually is the chemistry that holds individual chunks of greater and smaller masses of the matter cosmos tightly griped together no more no less. He believes that time travel may not be an adequate explanation to shorten the great distances one faces traveling between galaxy's and star systems. He explains how instantaneous space travel relative to the zero gravity fabric of space may not at all retain the element of gravity that calls to having to travel in the fourth dimension.

Gravity therefore is defined as an active moving balance of the universe's spin; galactic and planetary sphere orbitration and circularity mechanisms. In Quanta Physics Theory everything about the universe its existence and origin of existence are all now disclosed.

Gravitation

Gravitation, or gravity, is a natural phenomenon by which all physical bodies attract each other. It is most commonly experienced as the agent that gives weight to objects with mass and causes them to fall to the ground when dropped.

Gravitation is one of the four fundamental interactions of nature, along with electromagnetism, and the nuclear strong force and weak force. In modern physics, the phenomenon of gravitation is most accurately described by the general theory of relativity by Einstein, in which the phenomenon itself is a consequence of the curvature of space-time governing the motion of inertial objects. The simpler Newton's law of universal gravitation postulates the gravity force proportional to masses of interacting bodies and inversely proportional to the square of the distance between them. It provides an accurate approximation for most physical situations including calculations as critical as spacecraft trajectory.

From a cosmological perspective, gravitation causes dispersed matter to coalesce, and coalesce matter to remain intact, thus accounting for the existence of planets, stars, galaxies and most of the macroscopic objects in the universe. It is responsible for keeping the Earth and the other planets in their orbits around the Sun; for keeping the Moon in its orbit around the Earth; for the formation of tides; for natural convection, by which fluid flow occurs under the influence of a density gradient and gravity; for heating the interiors of forming stars and planets to very high temperatures; and for various other phenomena observed on Earth and throughout the universe

In 1916 Einstein sought to explain situations in which Newtonian physics might fail to deal successfully with phenomena, and in so doing proposed revolutionary changes in human concepts of time, space, and gravity.

The special theory of relativity was based on two main postulates: first, that the speed of light is constant for all observers; and second, that observers moving at constant speeds should be subject to the same physical laws. Following this logic, Einstein theorized that time must change according to the speed of a moving object relative to the frame of reference of an observer. Scientists have tested this theory through experimentation - proving, for example, that an atomic clock ticks more slowly when traveling at a high speed than it does when it is not moving. The essence of Einstein's paper was that both space and time are relative.

Rather than absolute, which was said to hold true in a special case, the absence of a gravitational field Relativity was a stunning concept at the time scientists all over the world debated the veracity of Einstein's famous equation, $E=mc^2$, which implied that matter and energy were equivalent and, more specifically, that a single particle of matter could be converted into a huge quantity of energy. However, since the special theory of relativity only held true in the absence of a gravitational field, Einstein strove for 11 more years to work gravity into his equations and discover how relativity might work generally as well.

According to the Theory of General Relativity, matter causes space to curve. It is posited that gravitation is not a force, as understood by Newtonian physics, but a curved field (an area of space under the influence of a force) in the space-time continuum that is actually created by the presence of mass. Einstein, theory was tested by measuring the deflection of starlight traveling near the sun; he correctly asserted that *light* deflection would be *twice* that expected by Newton's laws. This theory also explained why the light from stars in a strong gravitational field was closer to the red end of the spectrum than those in a weaker one.

CHAPTER ONE

THE UNIVERSE CELESTIAL GRAVITY AND THE THEORY OF FREE FALL

The Time It Takes To Travel"

Is time travel possible? Some think it is - but really is it? 4th dimension 'demeanor' measures by the threads of time as a dimension of itself. But space-time is based on the reality of the universes motion which is real and sit independent of itself being it is the universe itself. To say time is an actual dimension that interchanges within itself past, present and future actually means time is absolute and nothing in it can change. So time travel is impossible. (It either is(present) was(past) or never will be (future) because these aspects about reality and the universe are absolute. They are either present or they are not. Time travel would then have to come from the future earth and it doesn't exist at all except theoretically.

The question at hand is asked "What is Gravity"? The following question arrives "How does it work? To some persons of importance on this planet magistrate like Queen Elizabeth was very puzzles about trying to answer this question. In 1667 the queen assigned her imperialist Isaac Newton to research out answers to the mystery about gravitation. She asks Isaac "why do the apple fall to the ground'? The question should have read "Why do the apple fall from the tree" as a result for his task Isaac Newton discovered a mathematical sum that seemed consistent for answering the question about earth gravity (a truthful and correct answer should have read 'The apple falls because the apple has reached maturity raised from the planets natural elements – my queen).

Isaac measured the velocity and action of free falling objects. He discovered that no matter what the weight or size of an object mass – all masses seemed to free fall at 9.8 meters per second no matter what. The only indifference to this consistency of free fall he called 'attraction" or objects falling freely towards the earth's surface was the inconsistency of added force acting behind the free fall that even air diminishes away the higher you are.

No matter what – inconsistency was added by a push or acceleration overwhelming direction and pushing force behind an object in motion and changed the consistency of the objects free fall adding an additional pushing force to a freefalling object falling towards the planet's surface actually created a maneuvering passiveness to form by the freedom of the objects free-fall mobility. An object aviates by free fall passing through the earth's atmosphere measured only by the earth's atmospheric gravity dome or gravity ring. The part of the planet's atmosphere or bubble concealed by the planets momentum impressed due to the planetary orbital cycle mobilized within the fabric space grid plate.

As the earth rotates it is pressed physically against the fabric space grid. Because of the universe orbitration energy all galaxies as well as planets and star spheres embed themselves pressed and engraved against the space fabric itself. Because of the existence of matter formed by the big bang event that happens in all universe birth events throughout empty space the dark fabric by which the planetary matter laying relevant to this first beginning event is what generally mobilized a material space grid. Everything that exists in the universe was made in and from it thereby lies within its grid age.

While the universe was set in motion by the first big bang even impression of the first universe what followed was a good weightage of galactic and solar star system little bangs. All these following little big bangs all resulted by the same means as the first big bang in nature. The beginning big bang event started the grid motion in the dark element started by the inconsistency of what we might try and measure as immeasurability. While the massive first universal event mobilized an orbitrating cycle between matter and the dark element of space informing it into a grid smaller and shorter little bangs reformed the first event creating new and deeper grid plate age inside the newly formed universe. Inside these smaller galactic bangs even smaller and denser tiny bangs proceeded. Over the immeasurable time existence preceding us after the big bang event ordainments like Christmas lights sometimes proceed us outside are planets embedded space grid opening and allow us to observe what are called supernova events. Planetary stars and or possibly a system of stars or planets appearing from nowhere amongst the planetary observation we have already mapped of our solar system.

As you can see – the space grid gravitation element along with planetary gravity like here on earth measure and physically appear quite indifferent and they are.

Isaac Newton researched and deciphered the thread of object free fall passing through the earth's atmosphere at 9.8 meters per second. He also maintained that objects in free fall towards the earth's surface were an attraction. He led on to say that the attraction was formed by and objects mass energy that is weighed inside the molecular realm of all matter. He deciphered that in the same way a magnetic attracts other entities objects falling towards the earth's surface are attracted in the same way. It was later that Albert Einstein pursued the theory of gravity and assumed the idea that this same energy mass in dominate objects would built infinite mass in acceleration. That a moving object traveling at close to the speed of light could maneuver no faster due to the lack of energy. Relative to special and general relativity research – the speed of light was constant. Nothing could maneuver faster.

Quantum Superposition

Quantum entanglement is a physical phenomenon that occurs when particles such as photons, electrons, molecules as large as Bucky balls, and even small diamonds interact and then become separated, with the type of interaction such that each resulting member of a pair is properly described by the same quantum mechanical description (state), which is indefinite in terms of important factors such as position, momentum, spin, polarization, etc.

Quantum entanglement is a form of quantum superposition when a measurement is made and it causes one member of a pair to take on a definite value (e.g., clockwise spin), the other member of this entangled pair will at any subsequent time be found to have taken the appropriately correlated value (counterclockwise spin). Thus, there is a correlation between the results of measurements performed on entangled pairs, and this correlation is observed even though the entangled pair may have been separated by arbitrarily large distances. In quantum entanglement, part of the transfer happens instantaneously. Repeated experiments have verified that this works even when the measurements are performed more quickly than light could travel between the sites of measurement: there is no slower-than-light influence that can pass between the entangled particles. Recent experiments have shown that this transfer occurs at least 10,000 times faster than the speed of light, which does not remove the possibility of it being an instantaneous phenomenon, but only sets an active lower speed limit. (Quanta Physics Theory – *"Instantaneous Space Travel Acceleration"* read 'The Fabric of Space Gravity" 2013)

This behavior is consistent with quantum-mechanical theory, has been demonstrated experimentally, and is an area of extremely active research by the physics community. However there is some heated debate about whether a possible classical underlying mechanism could explain why this correlation occurs instantaneously even when the separation distance is large. The difference in opinion derives from espousal of various interpretations of quantum mechanics.

Research into quantum entanglement was initiated by a 1935 paper by Albert Einstein, Boris Podolsky, and Nathan Rosen describing the EPR paradox and several papers by Erwin Schrödinger shortly thereafter. Although these first studies focused on the counterintuitive properties of entanglement, with the aim of criticizing quantum mechanics, eventually entanglement was verified experimentally, and recognized as a valid, fundamental feature of quantum mechanics. The focus of the research has now changed to its utilization as a resource for communication and computation.

The counterintuitive predictions of quantum mechanics about strongly correlated systems were first discussed by Albert Einstein in 1935, in a joint paper with Boris Podolsky and Nathan Rosen. In this study, they formulated the EPR paradox (Einstein, Podolsky, Rosen paradox), a thought experiment that attempted to show that quantum mechanical theory was incomplete. They wrote: "We are thus forced to conclude that the quantum-mechanical description of physical reality given by wave functions is not complete."

Like Einstein, Schrödinger was dissatisfied with the concept of entanglement, because it seemed to violate the speed limit on the transmission of information implicit in the theory of relativity. Einstein later famously derided entanglement as "spukhafte Fernwirkung" or "spooky action at a distance."

It is by purpose and composition of this literature about the "f" force – of gravity and acceleration that the velocity of light is NOT constant. As you continue to read this book about earth gravitation and the universal space field grid which all material deities reside on – I explain a side about the gravity force which opens the passage wormholes not only to faster than light space travel but also – the only manner by which astronomical space field distances between planets and galaxies can be achieved. How time travel passing through the fourth dimension may not exist at all bending space as like a huge planet cannot even maneuver to how the speed of light may only exist with alignment to the density of the realm that may have to exist with it and why the space grid may not be pent ratable by any means. Being things to a more reasonable conclusion we should view the outer space grid formed as a rotating field grid which are planets, stars and galaxies reside on as they orbit and rotate on. Laying on top of Dirac's sea disk which acts as a grid mechanism operating as solar system, galaxies, Nebulas and stars systems the mechanism grids mobility relays independent to the embedded system of debris that reside its dominion.

Planetary matter occupying specific sectors on the space grid, above Dirac's sea grid sector lays a mist of open empty freedom space. Space that retains no ties to dominated existing occupied space grid mobilized by the existing matter residing its plate activity a cold accusable mist manifestation exist above the grid. Like quanta rays particles or light that only travels above the space grid activity the deep valleys and rivers that lay between the occupied planetary realms and mountains from which celestial deity's lay positioned throughout the universe , a space craft traveling at great propulsion speeds through the freedom dwelling that plays atop planetary Dirac Grid fabrications. With this disruption about the freedom space grid along came the reality of the ideas of time travel and short cut wormholes the idea that a direct slash into the Dirac grid plate would assure the ability to travel a shorter distance that following the mounted grid. A space ship like a boat called a submarine traveling below sea level – depending how deep a closer planet or star may exist traveling through the grid may have its advantages. Trying to manipulate the cold freedom cosmos field freedom where physicists talk about curving an atmospheric freedom atmosphere to create a maneuvering wormhole believing it will take particles sent into it backwards through time. The physical belief that there exist an extremely deep biological clock that can be manifested by a depth of velocity such existence could only be entered by intense overwhelming velocity shrinking space quantities fabric stringer fabric elements that allow manipulation of time. The same time and extreme density drama that allowed deep faster than light maneuver after the big bang event 13 billion years ago that bends and curved the dark space grid but not ripping it. It seems that anything made from a mother lode will not have the overwhelming strength to cause its breathier.

So the question about what gravity is still not really answered. A falling object descends towards the earth's surface at 9.8 meters per second. The earth elliptical passage crest line while the planet's atmosphere due to its oceans forms from it humidity water and rain. Clouds form above caught inside the planet's atmosphere with little means to escape due to the planets cycle continuum where thunder and lightning engage by the abundance of cloud vapor and electrical sparks layered into thunder bolts that stream towards a well vaporized surface point created by the presence of allusive sparks ascending throughout our planet's surface atmosphere. The earth rotates at 18.5 miles a second meaning that for every second of time that passes 18 and a half miles pass along the earth's equator.

All the elements I have just mentioned are what layer the atmosphere the lays between open empty space at the earth's gravity crest line and its ground surface. Due to the change in the seasons and atmosphere heat and cold vapors created on the opposing side of our planet at nighttime the different cycles our solar system orbits in its elliptical curvature around our sun that also creates different solar wavelengths heat and frost on the earth's surface great winds pass through the mountain terrain that a person can feel as a great invisible force of physical activity inside the atmosphere.

The biological turning of the planets journey around our sun passes measured in seconds compared to measuring the velocity of ground momentum we at the planet's surface measure in minutes and hours. Wheat we measure to be friction seems more so to be the activity due to the lack of constant force. One engages a propulsion force at the gas pedal of an automobile and calls it friction because the car slows down. Friction deemed as an electrical terminology holds little weight when it is actually weight that slows the automobile not any electrical drag. Objector force pushing against an extreme type opposing force measured in seconds on a clock seems more the reason a propelled machine would slow down relevant to orbital cycles measured by minutes band hours here on the planet's surface.

Planet gravity might be viewed in the same manner. A slow drifting object in orbital free fall is controlled by the dramatic element pressure force the planet creates as it orbits and rotates at extreme universal elliptical striving forces. The elements measure to be full atmosphere pressure a passive falling object renders through as it decelerates out of slow free fall. Each second the planet rotates 18.5 miles in its elliptical path – measured in the same measure of the second – the object falls 9.8 meters. As the second 18.5 miles the earth turns, the objects free fall velocity doubles to 19.6 meters and so on until the weight of the object free fall reaches terminal velocity. Terminal velocity meaning a velocity measured equal to the objects weight and the free fall of the object becomes terminate and will gain little to no more speed as it falls.

When we review and include the facts of the elements and pressure of the planet's atmosphere to explain gravity it's a lot different than just believing energy mass to be the source of our force of free fall. In quanta Physics Theory energy mass is what pods matter main chunks together that separate from main object core material – like pieces separated at the big bang 27 billion years ago. The center of mass holds the balance of allusive pieces existing separate from other chunks of matter and of different chemistry. Reasoning the facts and balance of terminal velocity energy plays no role to Newtonian attraction of a falling object – attraction is measured by positive and negative values and electrical terminologies having essence to magnetism. Einsteinium gravity steers to energy mass towards that attraction even through his celestial gravity terminology reframes a push and pull gravity force. Dominion stars create deep curvatures and bend space as the universe as a whole stretches the space fabric outwards towards the universe's outer rim. In this manner a balance occurs. More so – attraction in space is explained more as a free fall space curvature or stretching of space to a fabric that cannot rip when we research its matrix'.

The nature of his theory on gravity is formatted on the earth's natural elements air, earth, wind, heat and the nature that is enclosed on our planet's surface in space it is The Dark Element. Earlier physicists have defined the gravity force as an application of energy but a weak force which mass centralizes gravity as the center of mass in all objects. Even though it is the matter energy mass that everything is held together with at the molecular level – Kawecki believes it is not the cause of gravity itself. More so he entitles the gravity force as applied activity of the earth's natural surface elements and has explained the gravity force with the activity of nature itself rather than a weak force natured by energy mass.

Kawecki *laterally opens the door* to define the gravity force on the foundation of nature not energy and deciphers the equations and measurements that precede his theory. Why does falling objects no matter its initial weight or size all fall at a specific speed towards the earth's planet surface? What causes the action that precedes free fall and by what means does it happen? And finally – if the gravity force celestial or inner planetary atmosphere is not measured by the cause and effect of energy or energy mass that repels than what is it? In this book he will answer all these unknown questions about *"The Path of Gravity"*.

When we talk about gravity we are not talking about the fundamentals of energy and mass but more so is the activity of zero point energy presence and surface weight. Space is different. We are talking about how *similar* or like material entities called celestials such as planets and stars repel forming the distance between them in space present on the fabric of space that impresses them into deep holes. In space yes it's the presence of planetary energy in the elements that commands the divide. But in the interior elements of a planet's atmosphere there exists the *'ladder of resistance'* or an entanglement of free fall of an objects weight and acceleration as it falls. In the thinner upper atmosphere an object in free fall gains acceleration in layers through the atmosphere as the planet's atmosphere pressure increases making the object faster at variable speeds as it follows a curving free fall pattern towards the planet ground surface. Nothing travels straight down to the earth's surface.

The problem I have with ""G"" is in space similar matter (planets) according to physics repel and do not attract as today's modern physicists have been lead to believe over the years. Like a magnetic north pole and another north magnetic pole...they repel. According to the big bang theory if all was in advance a singularity bang universal, galactic or otherwise then these similar worlds of common ancestry matter ...*repel*... and don't attract wouldn't you agree ?

If this is the case than - a planetary gravity field "TORUS" otherwise...pushes away objects as in universe expansion theory explaining the pushing away expansion. Gravity thus on earth in this sense free falls at 9.8 meters per second compared with 18.5 miles per second it's rotational orbit...this is a free fall elemental falling object ...wouldn't you agree?

Do wormholes exist or can they be made using the energy capacity of a star? Even a star cannot penetrate the fabric planets and galaxy's themselves shelve themselves on in space. To think that breaking through the fabric of space would be possible lays the end of the wormhole travel short-cut theory in special relativity. It makes one believe whether or not Albert Einstein really believed that such things were special at all?

Rodney Kawecki tries to answer all these questions about gravity and more. This book will enlighten and supersede the most interested reader of advance cosmogony. Gravitation has at reach or range to infinity. However, it is the weakest of the fundamental forces. The gravitational strength is only $6*10{-}39$ of the strength of the strongest nuclear force.

Note: $10{-}39$ equals $1/1039$, where 1039 is 1 followed by 39 zeros. That is a very small number.

The strength of the gravitational force decreases as the square of the separation between two objects, as does the electromagnetic force. Although the gravitational force is much smaller than the other fundamental forces, it's impact concerns objects of large mass, such as planets and stars. Gravitation is what keeps the Earth and other planets in orbit around the Sun, as well as the Universe. (Lift-Vector (repel) $+C^2=G^3$)3D.

The Fabric of Space

Albert Einstein states space is a fabric and it is the curving of the space created by the weight of an object it sits in, that creates gravity. But if space is a fabric, shouldn't it have some measure of tensile strength? And if it does, the then the fabric of space itself is subject to entropy. There are parts of universe that are very dense, center of the galaxy for example. So naturally space is stretched downwards towards the center of the galaxy.

Einstein was right, again. Satellites that have been pulled slightly off their orbits show that the Earth is indeed twisting the fabric of space-time as it rotates, scientists said. They said their findings are the first to directly measure and prove an important aspect of Albert Einstein's General Theory of Relativity: that a rotating body warps and twists the "fabric" that combines the three dimensions of space and the fourth dimension of time.

"As the Earth turns, it is actually twisting space-time with it. Near Earth, the twisting is greater," "This twisting of space-time, which is also referred to as frame-dragging, has never been directly observed before, "Time and space, according to Einstein's theories of relativity, are woven together, forming a four-dimensional fabric called "space-time." The tremendous mass of Earth dimples this fabric, much like a heavy person sitting in the middle of a trampoline. Gravity, says Einstein, is simply the motion of objects following the curvaceous lines of the dimple.

According to Einstein, living in the universe is like living on a huge piece of soft elastic rubber. Space-time is a medium that has shape and form. Objects within this medium can flex and twist it. Every object in the universe pulls on the space around it, drawing the fabric of space-time toward its center. The more massive the object, the more it pulls. The amount of pull exerted by an object on the universal fabric is its gravitational force. So the apple falls to Earth because Earth has warped space-time in such a way that the apple must move toward Earth's center. More massive planets create a deeper warp, imparting a faster acceleration to objects that wander past.

Albert Einstein's theory of general relativity tells us that gravity is a curve in the fourth dimension of space and time -- and there's proof to back him up. What causes the curve is mass. Seriously weighty objects can bend the fabric of space-time. It explains why the planets orbit around the sun. The sun is so incredibly massive it essentially bends the space around it, pulling into orbit lesser objects (like planets) nearby. Similarly, with enough mass an object can even cause an otherwise straight beam of light to curve. In astronomy, that's called gravitational lensing.

Time is not immune to the effects of gravity either. It passes more quickly the less gravity there is, a phenomenon known as gravitational time. At the center of the Milky Way Galaxy seems like the last place to form a new planet, inhospitable and violent even. Stars crowd each other, whizzing through space like cars on a rush hour freeway while supernova explosions blast out shock waves and bathe the region in intense radiation. The very fabric of space is twisted and warped by a supermassive black hole's gravitational forces.

Geodesics: Universe gravity acts proportionate with each and every single planetary sphere and set of solar systematical sphere stars and galaxy's that in turn a fabric grid plate. According to Einstein a planet acts independent of itself as it rotates curving the surrounding space around it. The problem about Einstein physics is that celestial bodies don't attract they repel because they are formally of the same matter deity from forming out of the big bang. If this is the case, planetary bodies repel and don't attract. Accordingly a planet twists the present fabric space-time because it rotates not that the fabric makes it. It rotates due to the resistance between the planets and stars nearby that are also caught surrounding the dominion star our sun.

Because the planets, stars and galaxy's repel this is what causes them to twist. It is the weight of the celestial sphere that embeds the sphere in the fabric grid and it is the universe's spin that increase how deep in the space grid it retains. As the dominion star repels a planet within its grid its pushed deeper into the space warp age. As the universe spins it is impressed deeper into the fabric. Maintaining a proportionate mechanism of spheres and planets twisting occurs because of the greater dominions grip on the planet pushing it into a rotational spin axis maintained by the universe itself and its motion. The planetary cycles of the spheres throughout maintain a constant magnitude and velocity due to the constant orbit and rotation of the universe. Twisting space is not independent to a single planetary action it is universal and is what we call 'universal gravity".

In November of 1919, at the age of 40, Albert Einstein became an overnight celebrity, thanks to a solar eclipse. An experiment had confirmed that light rays from distant stars were deflected by the gravity of the sun in just the amount he had predicted in his theory of gravity, general relativity. General relativity was the first major new theory of gravity since Isaac Newton's more than 250 years earlier.

Einstein became a hero, and the myth-building began. Headlines appeared in newspapers all over the world.

On November 8, 1919, for example, the London Times had an article headlined: "The Revolution in Science/Einstein Versus Newton." Two days later, The New York Times' headlines read: "Lights All Askew in the Heavens/Men of Science More Or Less Ago over Results of Eclipse Observations/Einstein Theory Triumphs." The planet was exhausted from World War I, eager for some sign of humankind's nobility, and suddenly here was a modest scientific genius, seemingly interested only in pure intellectual pursuits.

The Essence of Gravity

What was general relativity? Einstein's earlier theory of time and space, special relativity, proposed that distance and time are not absolute. The ticking rate of a clock depends on the motion of the observer of that clock; likewise for the length of a "yardstick." Published in 1915, general relativity proposed that gravity, as well as motion, can affect the intervals of time and of space. The key idea of general relativity, called the equivalence principle, is that gravity pulling in one direction is completely equivalent to acceleration in the opposite direction. A car accelerating forwards *feels* just like sideways gravity pushing you back against your seat. An elevator accelerating upwards feels just like gravity pushing you into the floor.

The only problem *Quanta Physics* expert Rodney Kawecki has with general relativity is that the first car got there first? The understanding of gravity is that it is universal. Meaning that gravity on a planet experiences change due to the elements and pushing outwards of sphere energy in the essence of Einstein's planetary gravity. The bottom line is that planetary inference of its elements and energy change the density magnitude of the gravity force present from being universal by the adding of planetary forces acting within the gravity force field whereas space does not and it is universal in both instances. An object at rest experiences inertial because its acceleration is erected from its maneuver from being at rest state position or acceleration whereas in space there exists no external forces applying maneuverability except the fabric of the space grid itself.

CHAPTER TWO

EARTH'S GRAVITY

What is earth gravity? It was best explained by the queen of England back in the year 1637, when she asked Isaac Newton the famous question. Why does the apple fall off the apple tree to ground Isaac? The question at the time gave reason to Newton to research the aspects of the queens' question as Newton would later best answer the queen's question with the phrase "The apple falls from the apple tree due to a force called gravity".

It was later that Newton researched more on the idea about gravity that he later deciphered a free fall equation 9.8 meters per second. His research led to the idea that gravity concerning falling objects no matter what the size or mass of the object – large or small all surface matter seemed to fall to the planet's surface at the same rate of velocity. This was Newton's greatest discovery explaining the gravity force. Whether the falling object was a small marble or a large bowling ball both descended to the planet's surface at 9.8 meters each consecutive second of time.

This was the greatest stigma about the earth's gravity force that still stands today. But Newton was never able to define the reason for his mathematical decipherment for *the gravity force*. It wasn't until 1904 that a scientist namely Albert Einstein believed he resolved the mystery of what caused the gravity material mass polarity to different size mass descending object free fall. His theory relayed on the idea that all matter planetary celestial alike to even smaller surface objects retained a common interior force of energy that due to the dominion characteristics of a large celestial planet like earth smaller quantities or smaller surface objects all descended at the same velocity based on Newton's 9.8 meter per second free fall velocity because of their content of energy. Even though celestials in space do not fall they spin with the universe's rotation measured light speed. But Einstein did something else that quantized his theory about solving the definition of gravity. He asserted that at the celestial scale all the planets and stars were positioned due to their energy tent. Based on their size masses the planets and even the galaxies retained positioning as *ornamented* energy stars. The size mass of the planets energy throughout its sphere formed how far or how near the star positioned itself from another star. He also asserted and went along with Newton's theory that these sphere's and galaxies attracted one another. The real fact about space gravity is that the material star or planet sphere model to resists each other acting in a manner that allows the pushing away direction of an expanding universe. It also allows model planets spin alterations mobility or rotation cycles to be more easily understand lying throughout space on an attentive reflecting sensitivity fabric layered space grid.

He conceived that any and all chunks of atmospheric object pieces falling to the planet's surface acted parallel in descending towards the planet's surface because their energy at the molecular level retained a common massive energy which objects - due to then larger dominion of the planet's mass it allowed smaller objects to maneuver in parallel motion. That no matter the specific size of the falling objects because the earth's mass size dominated the spectrum atmosphere specific size mass of the ascending objects all descended at the same rate because their energy source parallel and acted relevant compared to the dominion mass energy of the planet as a whole.

In Quanta Physics Theory we take a new view to the explanations about gravity that Rodney Kawecki has referred to falling objects that have been agreed upon in textbook manner over the decades with the word free fall. As defined by gravity with the words free fall, by doing this Kawecki has tried to keep open the doorway to a new insight about earth gravity as well as universal space field gravity theory to explain his new terminology.

In review of the et forth explanations already explained in this chapter we can now look at the foundation of earth gravity and free fall of objects passing downwards through the planet's atmosphere or grid. To understand this more we can look at the facts about the earth's momentum in space itself the fact that the earth rotates in a circular motion every twenty four hours at 18.5 miles per second. This is not much to compare with gravity free fall velocity and the fact that 'terminal velocity 'occurs as the speed of an object becomes more than the weigh defined in its free fall towards the planet's surface. Unlike Newton's idea we should not get mixed up about his idea of attraction in the manner of which magnetic act in accord to positive and negative polarity. The earth is not the ground force for a positively changed falling object. Unlike Einstein's idea about gravity object free fall in this manner will not cause infinite mass calculations even at light speed the distance is too short and the probability illogical.

Yes it has been these facts that have confused theorist about the exact nature of gravity over the decades. In all like the perimeter terminology of planetary alignments in space and the space field grid disk plates planetary mechanism operates recycling irrelevant clock directions of star system without engaging the disk plate activity at the galactic scale that lay outside the inner galaxy. But gravity at the planetary scale seems to have to act accord with the rest of the universe. Meaning that as the earth rotates it also forms atmospheric layers of the deeper elements that reside within its atmosphere. As earth is a living planet it retains water, heat from the sun wind, rain and oxygen as well as a lot of other elements that are concealed in the planet's atmosphere protected gravity ring. A sphere of elliptical impression that push against the fabric space medium and what keeps the operating deities separate in the celestial cycles throughout space.

Isaac Newton's 9.8 meters per second in the second perception of time seems very small in comparison. But taking into consideration the planets actual rotation at 18.5 miles per second based on the twenty four hour cycle - the cosmological layers the planets rotational drift takes twists and spins the wind, air and oxygen that act as real invisible matter forces as a whole acts sufficient enough to create layers be which objects passing through the earth's atmosphere are slowed down as the debris descends towards the planet's surface. The common equation in this logic lays on the deciphered measurement is found in seconds of the physical biological clock that falling objects descend downwards in meters and the earth rotates in seconds with the celestial clock. Logic tells us that the layers in the atmosphere descending towards the planet's surface physically parallel with the spin of the earth's celestial plate that creates the density of the uniformity.

As we further decipher the mathematical possibilities of the celestial realm we can also agree upon the fact parallel descending objects of deeper mass quantities fall in accordance to the celestial material activity invisible forces reside. Celestial as well as object weight can be defined as the impressed engaging of mobile activity in a weightless free fall space until an action of motion is engaged weigh is least to exist at all. In 1963 as Neil Armstrong dropped a feather and lead hammer both at the same time in a one tenth single digit gravity seeing both objects descend towards the moons ground surface together and equally a long side each other. Is that to show that gravity acts as a single layer to all masses or is it that the moon's surface gravity measuring one tenth that of earths single g force acts as the reason or is it that the moon as a whole retains no element compoundable atmosphere by which specific layers and density mobility of these missing layers do not exist allowing both objects to not be affected at all by anything but close empty zero point space. An object almost floats as Armstrong leaps fifteen yards across the moon's surface. His golf ball today is still spinning around the moon satellite traveling at zero mass. Pursuing the outcome in the presence of the gravity force

CHAPTER THREE

ABOVE AND BELOW THE PLANET'S GRAVITY CRESTLINE

All planets, stars, galaxies space entities all retain what is called a "Gravity Ring". A gravity ring lies around the planet or star that acts as a division that physically divides the atmosphere density and elements that lay on either side of a planet. A planet exists in position in space or within a system of planets like our solar system. The gravity ring lays a distinctive field mass consisting on the inner planet's atmosphere dividing it from the outside atmosphere of empty space. The earth that lays within a specific space curvature it occupies orbiting within a system of planets orbiting our sun – it orbitrates around. From outside the solar system you can view the reflection of the solar systems sun rays reflecting and bounding off the interior planetary earth rock that lies inside. The earth rotates within a deep curvature of space the sun has trapped inside its larger space curvature or bending space fabric which in turn stretches out the suns sun curvature using the strength of the universe's orbit as it spins. It is the universes greatest interior extreme strength orbiting empty space wobbling massive planetary matter that lays inside its pod.

Planetary matter and energy has impressed its existence as a single rotating pod creating a continuous orbitration or spin with nothing in its way as it spins. Starting from what we call the big bang – celestial matter and stardust pushes its way which has caused the empty space fabric to curve in on itself using physical matter as a push driven force or propulsion continuum with nothing to slow it down except terminal velocity decipherable to its weight meta-forcibly a layer of the element. Because space itself is a weightless fabric impression all the matter that exist in it as a pod driven universe it impresses itself as the fabric walls building a material resistance against the fabric of the space field itself and matter in turn causing space to curve due to its unstable balance. Matter generated by the fabric of the space field itself cannot create a force greater than that it formed from it. So all the physical freefalling matter in it is subject will alternately curve and not pass through it nor can anything rip through it. This is a specific design of a divine creation that which resides in the physical universe as a natural continuum.

Outer space and planet gravitation

The question that lays in mystery about our universe I will logically attempt to answer at least partly in this small book about gravity. Physicists over the centuries have gazed towards the stars and wondered how they came to be. Biblical terminology called them lights. But in reality stars are nothing more than reflected light –

reflecting quanta particles off nearby planets throughout space.

Space itself is a vacuum. Meaning nothing exists but darkness except for light shining out and bouncing off other celestials. As a vacuum though – the space that lies between everything that exist out there is divided by this empty vacuum. Referring to exceeding the light speed constant maintained in 1904 by Albert Einstein scientist today have discovered that galaxy's seem to travel at a distance faster than light velocity. In fact galaxy's travels at speeds of close to 1.4 the velocity of light.

It is believed that if one traveled faster than the speed of light they would physically venture into the pass. As we make observations throughout outer space we see the deities as they existed far in their pass. When we measure distance especially astronomical distances we can only observe pictures photographed at the same time it takes light to travel. Most importantly a galaxy traveling faster than light would then be seen even further in its past age it existed than calculating its pictures by the speed light travels.

Light velocity because of these reasons is why a spacecraft couldn't travel faster than light without traveling into the past. Rodney Kawecki has tried to research more how the possibilities of space travel at greater velocities would be possible. One reason he resolved is that space as a vacuum of empty darken shadow between lighten stars and celestial deities retains no gravity like earth's atmosphere gravity by which would cause a space vessel to slow down as it accelerates. After reading his new insight in this book about gravity both planetary and interstellar space you will understand more about his theories.

Since traveling in the universes vacuum space field retains no counter acting invisible forces as earth's atmosphere does a vessel traveling at light speed in space would actually be traveling faster. Friction, gravity or whatever you may call it here on earth what slows moving objects does not exist so the power of propulsion is measured more in space than here on earth. Further in this book Kawecki explains a new theory about the gravity force never before researched nor explained before in history. More so you will be able to understand why galaxies as they do travel faster than light.

As I mentioned earlier in this chapter, it is not the dark element of the vacuum that controls the velocity of the universe's rotation − it is the interstellar planetary matter. Before matter formed the universe was just an empty space vacuum. Matter formed from out of the big bang event which modern physicists believe sums up the theory about matter in the universe. How the cosmic egg formed I will explain my thoughts about a little later in this book. But it is the force of the cosmic explosion which put action in the subtle empty allusion of dark empty space. The thought that other universes exist yet astronomically far off in the distance may allow us to rest assure that we are not alone in this.

The cosmic celestial sphere exploded forming a material substance from out of nowhere. This cosmic matter rushed at velocities much much faster than light making its impression in dark space. Since the universe as we know it today rotates at light speed steady in its orbit we must assume that the big bang was much greater. Forming out from the nowhere land – and only a substance of the dark element that which created it is much assuredly much stronger than the element it arrived out of. As the rush of the cosmic explosion pushed through this empty vacuum space it formed into a curvature because it is weaker than the dark space whom borne it.

This cosmic planetary matter pushed its way into an orbital rotation continuing its spin because there was no opposing invisible force to stop it. It continues to rotate to this day. The cosmic matter drug a long side to it the dark shallow dark element we call space pulling it with it as it moved through empty space. The cosmic matter could not break the dark element rip a hole into it or cause anything that would embed it free from its grip. As the planetary matter passed in its rotational orbit dragging the dark space within its grip is slowly over time slowed down in its velocity. That velocity today is called the speed of light.

From this view I ask you do you think it's possible to travel faster than the speed of light.

Logic tells us we can. Traveling at close to light velocity we travel through a vacuum equal or almost equal to the velocity the universe travels at today the velocity of light. This we can agree that are propulsion values having the capacity to travel close to light velocity – it's not more than a slight push that we overcompensate the light speed constant. Traveling with the current so to speak like rushing down a falling river – are vessel travels a long with the speed of the current. A propulsion value beyond that is created by the amount of force we can create beyond that point traveling the river bed no different than that of traveling through space.

We have now learned that traveling faster than light is possible – but how much faster?

It's called object free fall. Like earth's atmosphere objects traveling towards the earth's surface travel traveling against the earth's natural but invisible elements. In space we discovered that traveling beyond light speed means traveling into the space fabric's grade. The element itself it is made from. On earth we assume these interfering forces of elements are air, the weight of heat from the sun passing and reflecting over the planet as it rotates on a 24 hour cycle. As the universe rotates at light speed it drags a long with it space itself but also acts as a train of moving matter – matter which maintains a degree of constant density throughout its celestial body. Acting like a rushing tidal wave nothing to stop it – it pulls with it everything around and in it. It swings and turns in a vast cycler motion as the heavier celestial galaxies control a lot of its momentum. Inside these celestial bodies smaller deities orbit and take in the slack of the weaker space field that acts outside the galaxy it resides in.

Physicists used to think it was the swirling glass their hand swings that caused the water inside to turn, swing and orbit in a circle but it wasn't. The universe swings, stretches orbits and rotates by the massive ingredients inside the glass bubble. The bubble exist because it is the elliptical path of the planetary substance dragging it a long with it as it cannot indent, cut, rip or disco figure the dark element it resides inside – so it acts within an elliptical path and everything in it resides within its steady flow.

Earth gravity is not much different except for the invisible natural elements layering its atmosphere. As an object free falls through the sky – it is compensated to have to travel against the planetary elliptical passing of the earth's rotational cycle measured at 18.5 miles a second. The earths' atmosphere is layered in rotational spin that generates a constant flowing activity of its elements that generate greater in dense as the objects reaches closer to the planet's surface. At the surface the objects lays a rest. Object free fall is measured at 9.8 meters a second acting relative with the planets spin layering the planet's atmosphere in calculable seconds on the clock. An object will never fall straight down to the earth's surface. It will always take a widening route entering against the planets cycle. You can actually view earth gravity like a staircase with 9.8 meter (foot) rises every 9.8 feet. (Meters) We climb up the stairs a single foot a second. But in reality we are actually stepping down the staircase free fall beginning from the top.

The fifth element space can be layered, stretched, bent, impressed upon and traveled upon but everything will always stay within its grip.

QUANTA' PHYSICS THEORY

The idea in Quanta Physics Theory is that by doing the math faster Than Light theory intervenes with SRT and GRT and also Quantum Dynamics by Max Planck 1925. That such a flight travel of FTL does not include a mass less particle but that which involves weight relative to speed. As such 'light' becomes exempt in the dialogue. The theory to FTL mathematics include the intervention of 'e' energy and its potential source relative to the object in travel included with special relativity and Einstein's equation concerning the speed of light under the same circumstances. FTL is possible measured with the fact of material weight and empty weightless space. Relativity is a theory asserting gravitation. Empty space...has no gravity. The theory of special relativity and the light constant become exempt in the equation for FTL because of this. The weight complexity - becomes balance in the flight pattern at light speed when it's measured with the complexity of empty interstellar space or the vacuum.

Summed in the mathematical equation the velocity of atomic energy and speed relative with the vacuum the sum equals FTL. The equation is E=Mq2T-1second. e is energy equals m mass 'q' is 372,000 m/s minus one second. In the new theory and equation the vacuum assumes the density against the mass object at 10-34cm. 10-34cm (empty space) adverts to minus negative density to a positive number which assumes 'one' or the number '1'. The speed of relativity or light asserts the speed of 186,000 m/s. at 10+34cm. The difference is the vacuum figure relative to the positive gravitational planetary sphere the force of gravity at 9.8 m/s. To include this figure as a density we find it to be 10+34cm. The equation there for asserts that the speed of light in vacuum is measured from 10-34 to the positive number 1 and the acceleration to 186,000 m/s. From 10-34cm we then have 186,000 m/s. From zero or one to 10+34cm we have another 186,000 m/s. Together the equation sum is 'q' squared at 372,000 m/s. How does this new theory resolve the technological question of space travel in the fourth dimension? Special Relativity opens the door to the infinite future. *Quanta Physics* opens the back door.

The sum of this new theory and FTL superluminal space travel illustrates a new theory for superluminal FTL doctrine. The facts are real and so is the physical nature relative to the equation with the Universe. Not discovered until 2005. Quanta Physic is the new developing of the future of space exploration in scientific technology literature. Unlike any the theory in advance physics does Quanta Physics include the addition of its theory to intervene with any other in respect to its mathematical equations and physics description about the Universe? Quanta Physics is the newest physics theory that is believed to become the predecessor of *SRT, GRT and QED upon its release.*

QUANTA PHYSICS NEW FORMULA FOR INTERSTELLAR SPACE

Faster than light velocity Interstellar Space Travel is just another word for Time Travel. Because a Starship retaining the capacity to travel faster than light it travels into the past first before even thinking of the possibility to travel anywhere in time into the future. According to Albert Einstein and his 1904 theory on special relativity insist one travels into the future by traveling at light velocity new facts indicate that this is just not altogether true. You will learn why by reading this book.

A ship traveling faster than light speed travels into the past. It takes a lot more propulsion power for a Starship to travel into any planetary future than the speed of light. The figures below explain the reason why.

186,000 miles per <u>second</u>
18.5 miles per second
Earth_Rotation
subjected sum

1,000 gravitons

Looking at Albert Einstein's formula for traveling into earth's Infinite Future I discovered that the mathematics he used just don't pan-out as they said they do. Observing the figures - there exist no flight tie coordinates in the figures. It's just seconds, minutes, hours and days that correspond to a set of numbers multiplying time on a clock.

Resistance Not Attraction

Physical force continuum not energy

In the early 1600 Isaac Newton measured the force of gravity on earth. It measured 9.8 meter per second. Because a falling object fell in the direction of the planet surface he called this attraction. He also believed that it had to do with magnetism. Albert Einstein went on later and recognized the gravity force as an attraction made of energy. In both scenarios we discover that the gravity force is not made up of magnetism nor as Einstein predicted in 1900 neither is it energy.

As energy we can decipher that both the planet and the falling object repel each other possibly causing the bounce a falling object makes. He believed this to be a very weak force because of this. When we review the facts about the planet gravity force we discover that the object would gain considerable velocity in free fall if gravity were energy. But as Newton discovered object free fall ascends to the surface in second clock ticks fall intervals.

Each second of time the falling object gains an increase of 9.8 meters per second in intervals per second as it falls. As a weak force gravity energy there would exist no resistance on the object but there is. The free fall of the falling object gains velocity in specific intervals per second. It reaches terminal velocity when its weight and velocity weight becomes equal. It will increase velocity no further unless an exterior force is added to the free fall.

Resistance accurse when interrupted by the earth's elements which are very heavy when you look at it. The earth globe turns and the object mass are twisted into a free fall because of these elements. The planet is impressed against the dark element of empty dark space as it is pushed by the galaxy's rotational orbit accrued by the universe's spin itself. As the universe turns so does everything in it impressed itself against the fabric space field causing the planets to impress downwards into the fabric causing it to bend and curve into it. At the top of the earth's outer rim is where the weight of the planet due to these spin impressions and create celestial weight to planets and stars as well as the galaxies and impresses against the dark fabric pushing them into position. An open space curvature is formed with the celestial planet rotating at the bottom of the curvature depth. As the universe spins the angular momentum of the galaxy it also causes interior planets to spin as well. Though the interior galactic planets and stars are denser because they sit inside the galaxy their momentum is more persist then a running away galaxy that is divided by astronomical distances. This also allows galaxies to travel faster than the universe that the heavier matter pushes against in its elliptical path. This causes a massive interior pushing force dragging the dark fabric with it as it does. The galactic bodies' free fall as well as everything affected inside the universe.

On earth the free fall action of object free fall is pushed away as the force of the earth's rotation spin causing the surface elements invisible to the naked eye to create a heavy force itself thus forming a gravity ring or dome around the planet.

Is gravity an effect of controlled force or energy?

The Universe's Outer Rim

The universe was formed by a force which over time impressed what we call an elliptical path impression in the surface of the dark fabric of space. Like the universe everything inside it spins and rotates in the same way as the universe. The galaxies, planets, stars and celestial's we don't even yet know about talking about the earth and its gravity force we about the earth's outer rim. All planets, stars alike embed themselves with a deep impression forming a fabric space elliptical rim. Space as we know of it – was once believed to be a four dimensional deity or element retaining four dimensions or sides to it. Now – Quanta Physics has researched and discovered that it may not be that way. The dark element of space fabric what we call the vacuum seems more so to exist with celestial impressed at a surface deity that due to the universe's spin embeds a deep impression in it. The celestial sphere impresses the fabric and lays deep within a impressed hole which the planet resides. The planets impression causes resistance with the dark fabric using its celestial chemistry or deity elements to keep it separated from the fabric itself.

The dome like crest line we observe by the sun's reflection on it through telescopes is what causes the divide. The galaxy spins within the universe and the celestial planets inside the galaxy spin creating planetary disk plates like the one for earth we call the solar system. The solar system orbits pushing the planets in it into rotation within the fabric impression form what we call a solar year from the total force of the universe's spin as a whole. The earth rotates every twenty four hours as it cycles the sun every 365 days. The earth's impression in the fabric bends the space fabric as it is impressed in

Forming an outer rim observed as the fabric surface. As the universe spins and planets, stars and galaxies impress themselves in it they are pushed to the side we call angular momentum. Everything impressed in the universe dependent of its weight chemistry calibrates how deep a celestial is embedded in the dark fabric. Formally all celestials' galaxies alike all lay across a universal fabric surface formed by each entity celestial. When a ship travels journeying into space it follows a space curvature made by the planets deep impression. Like a free fall object trailing down towards the earth's surface – a ship journeying into deep space travels in the same type of curvature against the celestial elliptical curve.

Galaxies are the leading celestial debris in control of our universe. The dark fabric is nothing more than empty space that acts as a field for the debris manifesting itself out of the big bang bending and stretching the dark fabric matter assumed itself as head of the construction field. With nothing to stop it – the debris gained an infinites tic velocity causing the dark fabric to be dragged behind it. It swirled as it could break the fabric nor pass through it so it curved in on itself and formed into a strikeout medium of cosmic celestial substance. Its internal debris made up of gigantic separated chunks of celestial matter as the debris curved at an extreme velocity swirled and formed smaller embryos celestial spheres like galaxies, nebulas' and cosmic organic substances. It all manifested from the same chunk of the big bang so everything in the universe can be said to be similar. Most likely gigantic chunks of separated but independent chunks of divine chemistry – each sphere formed into a specific celestial formation. The Milky Way Galaxy we reside in is made up of a distinct material celestial material which its stars and planets revolved into specific types of natured substance planetary material. IN Quanta Physics we see our solar system as unique made up of nine planets which at its youngest beginning evolved from a very early evolutionary event of maturing from what *Quanta Physics* entitles as Cosmic Embryos Organics. Our early solar system matured in the same manner as the molecular atomic cycle. Each particle made in the molecular fabric material of celestial like our planet earth though retaining great suspension of energies different in molecular wavelength and polarities this interior substance making up the embryos in the same way managed to create a cycle similar to its atomic cycle. Slowly and in the same way as the universe created itself – its embryos' aged and expanded also.

Scientist today believes the universe is expanding and at a great rate faster than light. Like traveling the wavelength of a photon particle or 'quanta' quantum particle' there is no indifference having to do with changing time as some theorist decipher. Traveling at 1g the speed of light does not change time as we know it. Because we observe astronomical distances in galactic imaging it doesn't change the reality of journeying traveling a single light year as something more than the single year it is. The second hand on a clock does not change stop or slow down.

The universe's rotation cycle measured at light speed does not mean that traveling faster changes the matrix of time. Because we receive images from far off galaxies at a time in the galaxy observed past because it is so far away – we have not changed time. It's only that it took that might time for the photos to be received. We say that the picture we receive is a picture of the planet or galaxy ten years ago from its pass that a journey through time is possible if we can just travel as fast as the light the photo traveled to us from the celestial sighting. It is just that are technology to mobile lighten photograph through space is primitive.

Physicists believe the universe is expanding faster than the speed of light – yet light velocity is the universal speed limit persona'. It does seem that our universe has limitations when it comes to measuring its speed and momentum. Light velocity seems to be a classic number in the cosmos. Physicists say it's because quanta retains no mass or at least none we can detect of it. Yet it cannot pass through the fabric of space itself. Is that why light speed unique? Nothing can penetrate the space fabric – that binding material which matter seems to have manifested itself from. We come to asking ourselves 'How did matter any type of matter celestial debris alike come from'? I will try and answer this question in a divine spirit.

Late in the early twentieth century a man labeled the fifth element as 'Dirac's Sea". He assumed space acted in the manner as our planet oceans. So deep and so dark nothing can be observed. Is it possible he was on the right trail? Reviewing this idea about space we discover that the space field like our planets oceans retains a surface top. That deep down in its chambers life exist. If we never swam in the ocean we would probably never found that fish are in it. But space is still greatly different than the possibility about fish inhabit ting it. It is dark to the naked eye yet by reflection from light we can see some distinctions close to its upper surface.

The space fabric retains a surface in the same way but celestial material or spheres we call planets and as such must occupy and make its impression in it. But still we are asking the same question – where did the celestial come from? Like the earth's ocean down deep down inside where we can still not really observe exist material life like entities we don't know about. Way down deep in Dirac's Sea at its darkest depth where not even light is manifested and only darkest entails the depths life like entities living in pure darkest where light has never entailed so far down deep we can imagine invisible creatures may exist. On earth from depths much less measured than what space may counter – a research team reached the greatest depth where light had never surpassed and by luck in a reflection at a higher depth an invisible cock roach was discovered. They caught it inside a clear glass when later at top side they were able to observe it. They used light to reflect its movements in the glass.

The fifth element we entitle as the space field takes on a lot of the same characteristics I just told you about. The existence of material entities lying in the deep shadows where not even light can manifest itself. But space is more than that – a lot more. The celestials we see in the night sky that even exist during the day but are hidden by the sun's reflection of our planet all came from somewhere. Reviewing this situation we have to also look at the idea that a dark element celestial surface in the fabric doesn't exist until a celestial is present. But again where did these celestial deities come from?

To jump into enlightenment I will say that everything made in the darkest does not manifest itself until it comes to light. A cosmic egg or embryus sphere reaches a specific maturity within the depth of *Dirac's Sea* until it can no longer stay in it. The embryus has to come into the light. The embryus surfaces the grid and manifests itself in a chronic transformation. As an embryus it changes its physical cosmic bio--composition and tries maintaining its maturity from growing out from the Dirac's Sea into a second new stage maturing into a new physical bio-composition transformation it how fights to survive outside the fifth element matrix of existence it cannot travel back through the transformation from the deep depths of cold darkness manifesting into a cosmic egg. It now tries to maintain itself with its new entangling chemistry elements. Energy that manifested the embryus from its depths to maturity now entangle independently within itself. Cold freezing elements allusive free energies traveling freely through the empty space field begin to act on the new celestial's physical presence changing its physical composition. The cosmic egg fights to retain power into energy and as it succeeds it grows stronger and politically mature. Its power rises as it impresses itself against the fabric that borne it for so long before releasing it out from its depth. The cosmic celestial demented energy changes into a frantic cycle illuminating the bliss for power as it embeds itself deeper and deeper within the fabric's skin surface.

Finally it reaches a critical density point and bang it explodes above the darkness. Great gigantic chunks of mold blistering cosmic matter shadows and passes through the dark fabric it has raised above from. Pieces break off and smaller galactic embryos' celestials are formed on the wayside. The new cosmic celestial now aluminates great suns and planets broken separated chunks of molding masses fall independent to himself by the force of separation the dominion stars like the suns try to attach while outside the gigantic galaxy the force of the universe spins with great force.

Inside the galaxy's smaller systems are formed like our solar system. Life from the darkness rises when a planet retains the true and balanced chemistry for survival. The cosmic egg has fallen from being pushed from darkness and found a place deep in the grip of the divine fabric that manifests everything that exists throughout its cosmic touch.

Attraction, energy or free falls

Earlier physicists believed it was attraction that earth gravity stimulated a pull to the planet's surface traveling towards the ground. In 1904 Albert Einstein believed everything retains molecular energy particles and that it was these particles that attracted a falling object based on the falling objects mass size. Later it was discovered that energy repelled similar masses therefore kept objects separated. It was further found that an object deity retains a center of mass even when chunks of matter were put separated from the greater chunk.

When a small mass was separated from a larger mass they both gained center of mass energy poverties at the molecular scale particle energy that existed in al matter is what keep these chunks held together as a whole. So now we have discovered the molecular polarity that matter is kept solid.

Newton believed that it was attraction that objects' maintained a free fall direction to the planet's surface that the greater deity attracted the weaker object falling to the ground. In any or both instants if we look at the extremity of universal force meaning the universe orbiting space , galaxies pushing through the dark fabric at faster than light velocity formed after the bang. Little bangs followed and created galactic interior elliptical passage and rotation inside what should be considered a deep density realm Space between confined galactic bodies denser than those in open space. A planet appears out of nowhere and in the same way the cosmic big bang arrived from nowhere.

Forces like energy that holds the fabric of matter as a whole measured by a center of mass. Similar objects that repel each other seem not to allow for the earth's gravity force to qualify as an energy force. Attraction ascending out of a freefalling object seems not to qualify because it would mean that similar objects act in the same manner as a magnetic force again using energy for the cause.

When we review the criteria about gravity as a physical force see a dominion star like our sun which trapped its other nine planets in its fabric impression in space it was the universe's dominion orbitration that pulled the free fall planets trapped by the sun to rotate out into a balanced elliptical path of their own. The universe works in balance with physical forces that impress galaxies and planets with its rotational orbiting spin causing smaller deities like the galaxies to balance the interior planets and stars within their domain but which act in a denser atmosphere.

Quantum particles hold matter together, the staircase leading back from the heavens measure 9.8 meters a second dominated by the swirl force of the earth's rotation and planets orbital cycle held more so together and in balance by the universe in motion.

It seems easier to conclude that these invisible forces that include the quality of the extreme space fabric that bends, stretches and impresses planetary bodies on its surface allowing some embed deco within the skin. Formally we can conclude that the universe is well intact. A weakening of a surviving planet or star could cause a system of planets to collapse. This physical disaster could change a lot in that specific part of the sector. The planets and nearby stars would really realign themselves into other formations.

We see planets ads stars appearing out of nowhere called supernovas opening a passage from the other side of the fabric into a specific area between denser or even lesser dense areas in space. They pass through the dark fabric appearing between celestials as brighten deities which shine to an extreme for minutes and slow fade out like a passing candle. What is left behind is a molded planet or star. It was this link that made it possible for me to adventure new possibilities for a birth of a universe. The filtering of a cosmic egg which reaching maturity is pushed from the deeper realm. As whole the universe arrived in great detail as in a process of what I call 'similarity'. Similarity allows for the thought that universes as well as other existing planets or planetary system might inhabit similar species like humans. Multiple universe theory accepting the fact that universe's hold to the same connection of similarity that other universe's might exist retaining somewhat the same quantities' and natural deities as this universe. Twin universes or similar universe pods stretched across an infinite sky may explain the result of the many possible species that may exist if not in this universe in some other nearby.

It is this author's belief that he has opened new doorways to aspects about the universe never before analysis about the universe. Free fall force maneuver seem to work more so than the idea that it is a specific chemical interaction.

Time calibrations in any reflection of a space journey will not change the biological clock of the universe. Warp one traveling at the speed of light and in comparison warp two traveling at twice light velocity will only get you to your destination faster. It will not stop or slow down the cosmic clock. Comparing a photographs arrival speed is not a comparable entity compatible to trying to measure a change in the physical universe momentum. A spacecraft will not do the trick no matter what the velocity.

Starting a journey into open space can be viewed in the same manner that a house fly jumps into its travel course. It does not jump into a straight line but it does push towards a right angle all the time. Much in the same way a spacecraft launching a course into space does compensating for the planetary angle rotation motion. Motion occurs not directly but by a specific activity mode. The earth rotates at a 18.5 miles a second elliptical path orbit as the free fall earth's motion of free fall gravity measures in meters in the same time interval. The earth's rotation in its elliptical path around the sun impresses a deep and heavy natural activity of a steady specific motion in the spinning table. Some physicists try and measure specifics about light speed by this turn table spin hypothesis. At the higher end of the record album turn table the object travels slower taking longer to arrive its destination point of arrival. This activity is not measured by the area position of the freefalling object theoretically arriving to its destination faster it traveled nearer to the inner spin of the turn table. Time does not change slow down nor speed up as it engages in warp speeds from warp five to warp ten pushing closer to the shortest launch point. Really we are talking about a drain coning closer at its center. The universe spins naturally changing the physical depth of the cosmic drain that physicists left out the stretch to the universe's outer rim. It now acts as a flat disk in rotation spinning in orbitration there exist no three dimension bubble formed until velocities faster than light and to some scientist if the universe is expanding faster than 1g. Traveling at speeds greater than light only shorten the time it takes to arrive the destination. Time is only shortened by the velocity and time it takes to get there – geographic time does not change to any agree making actual time almost impossible. The universe or literal universe would have to stop and physically slow in the forwards direction of time and that my friend will not happen.

Why do galaxies travel faster than light up to speeds measured by the red shift of z<1. The idea that nothing travels faster than light seems to becoming contradictive. But planetary deciphered velocities seem also to hold a close to common bonding. According to physics it seems that any planetary body big bang, little bang erupts into an explosion measuring a specific density manifestation impression against the fabric grid. Once an embryus is formed and matures atop the grid surface it increases and uncontrollable density polarity which seems to measures close constant which the fabric holds accountable for. The fabric relative to cosmos field bangs happen at densities much greater than light or above an explosion density that dependent to the cosmos eggs physical chemistry – erupts than slows at close to light speed as it travels through an infinitely weak vacuum. Universe's measure and erupt relative to a specific frequency polarity force and extreme force impressing against the grid surface as it duplicates molecular chemistry particles that seem to erupt out from a most primitive infinitesimal yet infinitely impressed egg.

The idea here is that bangs that form denser galaxies or more allusive big bangs creating gigantic universe's measure and erupt relative to a specific frequency polarity force and extreme force is impressing it against the grid surface as it duplicates molecular chemistry particles that seem to erupt out from a most primitive infinitesimal yet infinitely impressed eggs. At the density of the embryus egg like close to all or most bangs throughout the universe a common denominator is close to light speed or a runner up galaxy velocity that measures faster.   When we review aspects about applied force we discover that propulsion and values of these mechanics' are what will take us to velocities faster than those limited to light speed as the propulsion value of a spacecraft vessels engine capacity *becomes* what controls the extreme force against the cold vapored space.

ESCAPE VELOCITY FROM PLANET EARTH

Escape velocity is defined to be the minimum velocity an object must have in order to escape the earth's surface field. That means traveling fast enough not to fall back to its surface. Escape velocity is about 7 miles per second or about 25,000 miles an hour. The object needs no additional force for the climb.

The escape velocity from the earth equals 11.2 km/s escape velocity from the moon is 2.38 km/s. Atmosphere friction (air resistance) a rocket traveling out from the earth's gravity well. 'Gravity Well' is the depth in a planets forced impression in the space fabric that causes a deep carnal bending of the space fabric. To travel out of this gravity well an object needs to attain

Earth's rotation has to do a lot with the escape velocity speed. The formula to decipher a planets escape velocity is v=sq. (2GM/r). M is the mass of the planet, G is the free fall velocity constant (9.8 meters per second, r the earth's radius and v is escape velocity. Nothing is thought about friction in escape velocity theory. The speed needed to escape the earth's *gravity well* is about 11,200 meters per second or 7 miles a second. Escape propulsion and the value of velocity is constant for such a run away from earth's gravity well. The ship has to reach outside the planets outer gravity fabric ring which the planet lays inside impressed in the fabric which the universe's spin keeps open at the upper fabric surface mainline. Since space acts as a fabric and science has proved this that open space lays about the flat surface grid.

This idea put a hold on passing through wormholes to make short-cuts in time. If the fabric is flat space caused by the universe's spin in response to a dominion stars deep dominated impression in the fabric – than where would the wormhole venture too?

It is the foundation of these types of question's relativity and other modern physics theories do not answer. It has been Quanta Physics lifelong idea to so.

ABOUT GRAVITY

In Special Relativity published works of Albert Einstein in 1904, Einstein asserts his idea about possible time travel on the perception in 'the earth would look as if it were and "*seem*" to be moving backwards in his late 1904 theory. Assertion of the word "seem" illustrates that it really isn't turning backwards but look as if. The truth of the matter is – is that the planet or system of planets in the circle would actually have to slow down and or begin and turn backwards for his theory to work. The idea that a spacecraft can travel fast enough (light speed) and the planets and or system of those planets will actually slow drag because the ship is traveling so fast and or equal to the universal disk plate, and make the celestial sphere's slow and reverse direction is a word to the impossible. I think that his idea about possible over exceed themselves when he talks about light speed and or any speed light. A system circular orbiting planet that acts amongst more planets around it in the same orbit will not change direction because of an off-shore spacecraft traveling at intimate speeds greater than light.

Time travel will have to discover a new and different route for changing the hands of time. Using wormholes in the same manner as the weight of a planet impressed against the dark fabric would have to rip through to meet on the other side nearby a destination. Reading my book introducing my theory about "Instantaneous Light Speed" engages space travel on the property of propulsion capacity and free fall in the Universe not on things that warren impossibility. So lays the can of wormholes.

Before the discovery of special relativity, Isaac Newton's theory of gravity was lacking the fundamental conclusion to what gravity actually is. Even though his theory can be used to predict highly accurate dimensions about how objects move under the influence of gravity, it offered no insight into what it really is. Newton accepted the existence of gravity and went on to develop equations that accurately described its 'effect'. But he never offered any insight to how gravity worked.

In 1907, Albert Einstein working in Switzerland created a new and central insight to issues he pondered at his desk at the patent office that eventually led him to the radically new theory on gravitation. He recognized that gravity and 'accelerated motion' were profoundly interwoven as a key. His insight revealed the idea that free-fall expressed in a falling elevator compartment, that the person in the elevator would not feel the 'effects' of accelerated motion from others without acceleration but who are assumed with gravity. He realized that being confined in a compartment changed the feel of force the compartment itself assumed if accelerated outside. That these 'affects' were found in up and down motion but not in inertia or vertical momentum.

His work finally granted him the recognition of his new work on "Special Relativity" and the Nobel Prize later. He discovered that the force you feel from a gravitational field or from accelerated motion was indistinguishable. Asserting these spectaculars about relativity we come to the foundation in Quanta Physics Theory that expresses the "gravity force" amongst the heavenly bodies as Newton did in his theory. The fact remains that Einstein's theory on gravity are quite indifferent than those of Newton's. Of course, one theory has to do with the planets earth and stellar gravitation and the other Einstein's theory of gravity with the planets and/or systems of planets and stars altogether. Are they different we will see? Quanta Physics Theorist have come together to discuss an interstellar space field retaining the same quantity powered deities as the gravity force expresses here on earth but do they in outer space? Intervening with the ideas Einstein asserts with the levitation of the compartment elevator with gravity he finally asserted in his completed theory that the heavenly bodies were calibrated in some way with dark matter or the fabric of space itself that holds the planets in place and in motion with adjoining other planets nearby and amongst them in space. As critical as it seems, as Einstein discovered how acceleration affects the mass of a body in motion, he regarded this discovery within the foundation that a body in motion asserts a particular amount of energy, thus affecting its potential velocity. Within this framework Quanta Physics theorist predict that whereas the planets and/or system of planets, stars and or clusters and galaxies in orbit which independently cause them to rotate due to this motion amongst them as a whole, they create a 'frequency of motion' relative to their matter-energy made up of specific waves within a sphere or object. With this in mind, we can assume that gravity is recognized based on 'energy' formulated in the bodies (stars) motion generating an energy-frequency. As other massive bodies are nearby 'matter-waves' in the body increase and/or decrease dependent of their individual size masses that generate as the body is in constant-motion. In other words, it is the planets 'matter-waves' that generate the particular wavelength or frequency of the sphere, that acts as the planets protection mechanism or 'attraction force' at a constant motion and frequency. Nearby planets of like matter and/or matter-waves create and adjust each sphere's specific energy amount that in turn acts to balance and position a system of planets in orbit as a group. The attraction of a larger more massive planet, star or galaxy in space is thus minimized to the distance a like matter-wave planetary sphere will become attracted to it. The story proceeding this amazement of Albert Einstein you can see how acts as a force that tie and keep chunks of matter celestial or bits and pieces together and what is now explained as the true nature of gravity both in empty space and which entangle planet invisible natured weight forces.

The free fall of an object within the earth's atmosphere is constant. The velocities of a freefalling object no matter what the size mass, width, thickness or chemistry accordingly and revealed by Newtonian physics is 9.8 meters per second. No matter the subjects' weight, size, chemistry or mass, a feather or a hammer they will all measure the same free falling deciphered speed. But the question is why? Why does a lead ball measured with the size mass or a small marble much smaller and much less in material weight – both falls through the planet's atmosphere at 6the same speed? This is the real question and I am going to try and scientifically address the answer to this phenomena.

When we think about 'gravity' we look at two specific possibilities in the free fall phenomena. (1) Object free fall within the earth's atmosphere where the planets invisible elements reside within the atmosphere of the earth itself act relative within a galactic dwelling we call the solar system and the galaxy evolves within the universe itself making all activity universal. It is the planets slow drag because it's confined within a solar system divided to itself on what is called a rotating disk. A rotating disk plate is the confined balance the earth evolves on empty space impressed by the it's evolving motion as it is embedded within the element of the space fabric allow. Amongst other planets acting on a larger disk impression made by the sun because the universe sets all planetary matter within the universe in an angular momentum acting as a cycle the planets orbit the sun on its disk plate impression within the large galaxy. In reference to all actuality – nine planets orbit our sun rolling full circle motion as it takes 24 hours what we call a full day, one rotation cycle.

(2) As a dark space field only lightened by the reflection of a burning sun that light reflects off other planetary bodies and be seen abroad empty dark space. The celestial bodies lay within atop deflectable dark empty space assumed a vacuum as its physical chemistry measures zero point energy/gravity squared ($-e/-g^2$). On earth objects in the field fall at a specific rate through empty space that is filled with planetary atmosphere upper surface elements. Velocity due to the planets motion, orbital spin and rotation are all that act accordingly and support a system free fall coordinates for the planet's atmosphere. A falling objects velocity and acceleration are measured in balance by the earths thrust and free motion accompanied by the planets physical rotation and cycle around the sun speeds. Accompanied chemical element weight of a single object or chunk of mass is obtained as the falling object gets closer and closer to the planet surface. Above where the element air, wind heat etc.., slowly diminish into a cold zero point gravity atmosphere that lays

outside our planets heavier inner atmosphere at the surface are chunk of matter retains a measurable material weight substance. In space an object lays motionless based on the dark element vacuum retaining no interrupting planetary affects or presents of invisible elements. An object relies on a physical interrupting force of some kind for its motion or momentum. What we call 'friction' in all actuality is really only the lack of invisible forces or lack of enduring force acting on the objects presence.

In 1904, Albert Einstein believed that empty space acted on the presence of an array of matter-energy occupying regions in space. The space field he named as a fabric was based on this observation and conclusion. He therefore assumed the interstellar vacuum of empty space to negative energy as did. Nikola Tesla 1856 – 1943) was an American inventor, electrical engineer, mechanical engineer, physicist, and futurist best known for his contributions to the design of the modern alternating current (AC) electricity supply system Tesla gained experience in telephony and electrical engineering before immigrating to the United States in 1884 and gained work for Thomas Edison. Two actors amongst the many from which the regions of what we know today as *empty infinite space* is just that – empty infinite space a dark element or dark void that acts in the presence of confirmative matter through which *cosmos galactic matter* or big bang gained unification.

CHAPTER FOUR

THE HISTORY AND WORLD OF THE BLUE PLANET TERRA

Earth is the third planet from the Sun, and the largest of the terrestrial planets in the Solar System in terms of diameter, mass and density. It is also referred to as the World, the Blue Planet, and Terra. Home to millions of species, including humans, Earth is the only place in the universe where life is known to exist. The planet formed 4.54 billion years ago, and life appeared on its surface within a billion years.

About 71% of the surface is covered with salt-water oceans, the remainder consisting of continents and islands; liquid water, necessary for all known life, is not known to exist on any other planet's surface. Earth's interior remains active, with a thick layer of relatively solid mantle, a liquid outer core that generates a magnetic field, and a solid iron inner core.

Earth interacts with other objects in outer space, including the Sun and the Moon. At present, Earth orbits the Sun once for every roughly 366.26 times it rotates about its axis. This length of time is a sidereal year, which is equal to 365.26 solar days. The Earth's axis of rotation is tilted 23.4° away from the perpendicular to its orbital plane, producing seasonal variations on the planet's surface with a period of one tropical year (365.24 solar days). Earth's only known natural satellite, the Moon, which began orbiting it about 4.53 billion years ago, provides ocean tides, stabilizes the axial tilt and gradually slows the planet's rotation. Between approximately 4.1 and 3.8 billion years ago, asteroid impacts during the Late Heavy Bombardment caused significant changes to the surface environment.

The Earth is surrounded by a blanket of air, which we call the atmosphere. It reaches near or over 600 kilometers (372 miles) from the surface of the Earth, so we are only able to see what occurs fairly close to the ground. Life on Earth is supported by the atmosphere, solar energy, and our planet's magnetic fields. The atmosphere absorbs the energy from the Sun, recycles water and other chemicals, and works with the electrical and magnetic forces to provide a moderate climate. The atmosphere also protects us from high-energy radiation and the frigid vacuum of space. An envelope of gas surrounding the Earth changes from the ground up. Four distinct layers have been identified using thermal characteristics (temperature changes), chemical composition, movement, and density.

The bombardment of Earth nearly 4 billion years ago by asteroids as large as Kansas would not have had the firepower to extinguish potential early life on the planet and may even have given it a boost, says a new University of Colorado at Boulder study. Impact evidence from lunar samples, meteorites and the pockmarked surfaces of the inner planets paints a picture of a violent environment in the solar system during the Hadean Eon 4.5 to 3.8 billion years ago, particularly through a cataclysmic event known as the Late Heavy Bombardment about 3.9 million years ago. Although many believe the bombardment would have sterilized Earth, the new study shows it would have melted only a fraction of Earth's crust, and that microbes could well have survived in subsurface habitats, insulated from the destruction.

"These new results push back the possible beginnings of life on Earth to well before the bombardment period 3.9 billion years ago "It opens up the possibility that life emerged as far back as 4.4 billion years ago, about the time the first oceans are thought to have formed" se physical evidence of Earth's early bombardment has been erased.

Meteorites and the pockmarked surfaces of the inner planets show evidence of a violent environment in the solar system during the Hadean Eon 4.5 to 3.8 billion years ago, particularly through a cataclysmic event known as the Late Heavy Bombardment about 3.9 million years ago. Although many believe the bombardment would have sterilized Earth, the new study shows it would have melted only a fraction of Earth's crust, and that microbes could well have survived in subsurface habitats, insulated from the destruction.

The possibility that life emerged as far back as 4.4 billion years ago, about the time the first oceans are thought to have formed" Because physical evidence of Earth's early bombardment has been erased by weathering and plate tectonics over the eons, the researchers used data from Apollo moon rocks, impact records from the moon, Mars and Mercury, and previous theoretical studies to build three-dimensional computer models that replicate the CU-Boulder researchers even cranked up the intensity of the asteroid barrage in their simulations by 10-fold -- an event that could have vaporized Earth's oceans. "Even under the most extreme conditions we imposed, Earth would not have been completely sterilized by the bombardment,"

Instead, hydrothermal vents may have provided sanctuaries for extreme, heat-loving microbes known as "hyper thermophilic bacteria" following bombardments. If life had not emerged by 3.9 billion years ago, such underground havens could still have provided a "crucible" for life's origin on Earth.

Researchers concluded subterranean microbes living at temperatures ranging from 175 degrees to 230 degrees Fahrenheit would have flourished during the Late Heavy Bombardment. The models indicate that underground habitats for such microbes increased in volume and duration as a result of the massive impacts. Some extreme microbial species on Earth today -- including so-called "unboilable bugs" discovered in

hydrothermal vents in Yellowstone National Park --
thrive at 250 F.

Geologic evidence suggests that life on Earth was
present at least 3.83 billion years ago.

The history of the Earth describes the most important
events and fundamental stages in the development of
the planet Earth from its formation 4.6 billion years ago
to the present day. Nearly all branches of natural
science have contributed to the understanding of the
main events of the Earth's past. The age of Earth is
approximately one-third of the age of the universe.

The earth was born a distilled planet until it would not
have been completed and sterilized by the
bombardment." The cataclysmic event known as the
Late Heavy Bombardment seems to physicists to be the
most modernized predictable theory about how life
began on Earth. More so it seems to be the most
acceptable theory. It tells a story about our planet which
adding possibly foreign alien element from space of new
and different debris allows the idea that stardust may
have been earth's missing link to its evolutionary space
period. We review the unique possibilities about how
life started and we picture a gigantic universe from
which are little blue planet resides inside a larger galaxy.
Darkness bellows enlightened specks of light in a dark
sky where on the other side of the planet the light from
our ingenious sun blinds our species hindsight of what
lays above us.

As mentioned earlier – the planet rotates at 18.6 miles a second. Free fall acceleration is measured inside the earth's gravity crest line that acts like a bubble surrounding the planet and measures free fall at .8 meters a second in the difference between them.

As mentioned earlier the earth's motion rotation velocity is 18.6 miles per second (not meters). Free fall 'acceleration' is measured at 9.8 meters per second (not miles) and relatively of what gains the heavier element in the atmosphere of our planet at its surface measured inside the earth's gravity crest line that acts like a bubble surrounding the planet and measures free fall at *meters a second* in the difference between them. Measured between the two equations act relative to one another based on they are clocked by the single second on the clock the difference in the earth's rotation per miles each second acting on the earth elements inside the earth's atmosphere cause the surface element mostly air, to ignite into an invisible weak force layering the planets inner atmosphere into a gradually gaining deeper tension and pressure as the atmosphere becomes more layered closer to the planets ground surface.

Understanding this concept about earth gravity as a process we should look at the original deep fabric of empty space itself. Empty space fabric lays everywhere even between atom molecular structures as it was mixed inside through the earliest universe's explosion called The Big Bang Event. Layered in space and what divides the celestial spheres abroad deep space measures zero point gravity in the comparison fabric that aligns and acts as the cause for positioning the planets and galaxy's in space. It is called The Fabric because all material deities, stars, planets, suns alike reside embedded within its grid.

Embedded deep within the fabric space rotating disk plates lay large celestial galaxies, planets and stars to as mall as tiny virtual particles popping in and out hiding in the fabric element. As the universe rotates as a motion material slow drag all material deities consisting of matter follow in a free mason elliptical space curvature dragging with the fabric itself. Impressed planetary matter push against the dark fabric causing it to twist and turn as mater paves its way into a consistent path through space. Orbiting an elliptical path matter builds up a celestial impression creating the turn-style mater into having physical concentrated weight properties. It is the dark element or existing empty fabric that sits adrift being dragged by the heavy galactic matter that was set in motion after the big bang event. Some physicists believe now that smaller bangs are what formed the already formed galactic matter and physically mixed with dark matter space billions of years later after the first cause.

We can view the fabric of space as a dark ocean that sits adrift as matter celestial objects like a boat in earth oceans impress its stillness into it. Likewise an object falling towards earth's surface it falls through the same embedded fabric yet natural shielding of earth elements like its air, wind, heat cold vapor act as layers caused by the planets rotation in the same manner as matter in space causes space to bend as an impressed disk plate. The subject matter is not that all different comparing the two especially when we observe empty space or dark space as the underlying source element of cause. In turn, the elements especially in empty space fabric cause the planets rotating against the universe's spin increasing its weight into its impression in the fabric forming celestial density levels to intervene the matter solidity.

The earth's atmosphere is not empty but filled layered elements creating heavy pressure. In the same way matter fills empty space into invisible impressed density space curvature. On earth on can surely feel the invisible layered air vapor in the atmosphere at the swing of a hand? Still the question is asked 'why do objects all fall at 9.8 meters a second Is it the earths influence acting against the dark fabric of space so unique it shields the planets elements into invisible layers that rotate in conjunction with the earth's spinning 18.5 miles a second which velocity forms a heavy atmosphere impressing these elements in the empty fabric that actually is everywhere even in the earth's atmosphere gravity ring that surrounds it. Remember in empty space the empty fabric is filled with planetary matter set in motion by the universe's spin – pulling dark matter space along its trail not space pulling matter. Planet matter likewise the physical presence of earth elements filling its atmosphere would also act the same way and be impressed in the fabric as well. Embedded in an elliptical path the atmosphere that builds its division against the fabric causing it to warp or twist is the same area the atmosphere resides forming its elements in heavier properties'. Remember it is only 'virtual particles' that swim in and out through the dark fabric not anything else known to us.

The earth's gravity elements are not permanent enough that they create a deep elliptical passage outset. As the earth rotates miles per second the stillness of the elements are stirred into a chronic allusive membrane we call the atmosphere. It is a fully graded atmosphere membrane from which protects the planet surface from space.

It might not seem like it, but air has weight. Anything with mass has weight, and we know air has mass because (for example) we can feel it when the wind blows. The total weight of the atmosphere exerts a pressure of about 14.7 pounds per square inch at sea level. You don't notice this weight, however, because you are used to it.

The atmosphere of Venus is about 90 times heavier than that on Earth.

The total weight of the atmosphere exerts a pressure of about 14.7 pounds per square inch at sea level. You don't notice this weight, however, because you are used to it. If you live in Denver, Colorado, which is at an elevation of about 5,000 feet, then about 85% of the atmosphere is above you, resulting in an air pressure of about 12.5 pounds per square inch. At the top of Mount Everest (over 29,000 feet), only 30% of the atmosphere lies above you, leaving an air pressure of only 4.4 pounds per square inch.

Lightning is produced in thunderstorms when liquid and ice particles above the freezing level collide, and build up large electrical fields in the clouds. Once these electric fields become large enough, a giant "spark" occurs between them (or between them and the ground) like static electricity, reducing the charge separation. The lightning spark can occur between clouds, between the cloud and air, or between the cloud and ground, cloud-to-ground lightning. Cloud membranes fall out of hindsight path and collide with another thunderstorm cloud usually thunderstorms are very much windy.

Twenty miles per second is quite fast for a planet velocity to rotate at. More so when we think about how the big bang event caused the mixture of empty space at the molecular level. The explosion event had to be very great and enormous to think that everything existing out there is from a point one singularity event.

Any object passing through the planet's atmosphere membrane undergoes falling through specific layering's slowly getting thinner at the top closest to outer space and more so heaviest at the bottom surface. Since the planet's atmosphere acts as a membrane as a whole we can speculate that the falling gravity layers are formed from the more intense membrane surface layers. Since the free fall of the object velocity increases each second of time it falls – than the degree of concentration in the layers intensify relative to the planet rotation velocity each second.

The free fall elliptical pattern acts as a staircase allowing the object to fall increasing velocity doubling it every second 9.8 meters. Finally dependent to its actual physical weight for example – the thrust of a mountainside waterfall rushing downwards towards the surface enlightened mist water drops spray to the wayside less massive than the bulk of water falling. The spray slowly diminishes away as it becomes lighter than the moistened air around it.

There exists a specific point of an objects free fall velocity where the object will no longer gain speed. It is when the objects actual weight meets equal to its allowed speed limit during free fall. Its velocity meets with a specific speed limit with the free fall and it no longer will gain velocity as it continues falling. Weight equals velocity per second called 'Terminal Velocity". Like the mist in the waterfall – the objects weight becomes equal with the air mass layers inside the membrane atmosphere. Depending on how large and heavy an object actually weighs will it meet terminal velocity during the free fall. An object weight can be greater than the heavy atmosphere of the earth membrane. Since the earth spin at 20 miles per second and quite fast for its atmosphere – it is very unlikely that any object will fall straight forward to its surface. Again, the objects weight should outweigh the concentrated air pressure of the elements in the planets membrane and would rely mostly on the objects free fall velocity if it were great or its actual weight mass as it falls.

The explanation about the gravity force and its affects render an allusive entre with energy. A lightning bolt is caused by two thunderclouds colliding into each other's elliptical passage creating a spark large enough to form into a bolt of lightning heavier than the atmosphere mass membrane ignites towards the ground.

Constituting 78.09% by volume of Earth's atmosphere nitrogen is a common element in the universe, estimated at about seventh in total abundance in our galaxy and the Solar System. It is synthesized by fusion of carbon and hydrogen in supernovae. Due to the volatility of elemental nitrogen and its common compounds with hydrogen and oxygen, nitrogen is far less common on the rocky planets of the inner Solar System, and it is a relatively rare element on Earth as a whole. However, as on Earth, nitrogen and its compounds occur commonly as gases in the atmospheres of planets and moons that have atmospheres.

Many industrially important compounds, such as ammonia, nitric acid, organic nitrates (propellants and explosives), and cyanides, contain nitrogen. The extremely strong bond in elemental nitrogen dominates nitrogen chemistry, causing difficulty for both organisms and industry in converting the N^2 into useful compounds, but at the same time causing release of large amounts of often useful energy when the compounds burn, explode, or decay back into nitrogen gas.

Nitrogen occurs in all organisms, primarily in amino acids (and thus proteins) and also in the nucleic acids (DNA and RNA). The human body contains about 3% by weight of nitrogen, the fourth most abundant element in the body after oxygen, carbon, and hydrogen. The nitrogen cycle describes movement of the element from the air, into the biosphere and organic compounds, then back into the atmosphere.

Molecular energy mass is a strong force which hold particle molecular structures intact as a whole or chunk of mass. Where energy masses related to matter retain a center of mass due to their content – it is unlikely that energy mass acts as a weak force sometimes related to the weak gravity force. Energy can be said to be tried into matter as an electrical terminological substance or mass rather than a gravity post datum.

It would also seem to matter that "Attraction" illustrated by Isaac Newton in 1667 known as Newtonian Physics may not at all be as true as presented to the public. That the larger bulk of multiple masses would be received faster than the lesser mass object but evidence shows otherwise. All objects no matter their physical size and nature all free fall at the same velocity 9.8 meters a second it is also shown that objects (asteroids) for instance gain weight as well as acceleration as they travel to the surface of a planet. The only other acceptable version to this theory about free fall is that if the object as an accelerated asteroid or comet traveling at a velocity un-parallel with earth's acting element pressure and traveled at rates per second in thousandths of miles per second than velocity intervenes as the controlling disorder and an object like an asteroid could free fall in a specifically measurable straight line to its target the earth's surface. This would be the free fall gravity accretion.

When we review the aspects of "Terminal Velocity "at the interstellar field rate of travel we resolve the indifference by discovering that it is terminal velocity that retains the sole speed of a traveling object through space. The weight and measures stem from that philosophy. Using Pluto as the targeted planet a traveling asteroid might shatter due to the planets resistance (gravity). According to Newton theory the asteroid is attracted to the planet illustrating attraction or a welcome to free falling assignments but as we review this Newtonian idea it retains little logic. Even Einstein's theory of energy mass gravity (star) like poles like in energy repel not attract. Postal to attraction the asteroid is not welcome to the planet surface. It travels by a retained velocity according to its mass-weight as it travels through zero energy (Z.P.E) atmosphere emptiness. As the asteroid reaches Pluto atmosphere its speed is counteracted upon by the planets resistance to its intrusion. Energy like the asteroid and planet chemistry repel not attract. As a positive energy type or due to the new planet's atmosphere which is very heavy as it creates a dome type protective atmosphere around the planet rock. If the planet atmosphere elements are strong enough not relative to its energy type chemistry the governing layers that stretches out towards zero point space might shatter the asteroid before it hits ground. If not its obtained governing weight will increase haven come out of terminal velocity when it arrived from space?

As we exceed into the matrix's of velocity in zero point gravity pace we are appointed with the sensation of a time matrix constant that everything in the universe or possibly different in another galaxy is predetermined by "TIME". A Pulsar Star impulse waves operate as a time interval based on a single second of time. Along with what Einstein derived as the 'universal constant" relative to velocity we cannot overlook the fact about time that the velocity is encountered by a constant of time as in the Pulsar Star activity. In this galaxy and perhaps the whole universe we are met by a matrix constant of time at second by second intervals that act as a base line for all formal matrix and energy content. Velocity in space changes to easier intervals to use seconds as a baseline instead of hours when we review the great distances of astronomical space travel. Propulsion values are met by the assessment of the engine power rather than the idea that "light" is the essence of the material spacecraft, it is not light we are talking about but the propulsion value to push a fruitarian spacecraft to the limits.

It is not a light beam that is pushing our ship through space but the propulsion capacity that exceeds in Quanta Physics Theory into a faster than light scenario. The value of using time on a matrix table is the ability to advance the impulse waves into units smaller than a second or greater for the value of interaction defenses. Is it that light is the constant for velocity or is it that time is the constant for advancements in both time and space?

Aside from the value grade for terminal velocity escape velocity is the speed an object must be given to escape from the Earth is 25,300 mph." The distance to space is small – space isn't very far away. You would have to get pretty close to Earth to be able to see the thin atmosphere surrounding our planet.

The reality is that there is no clear boundary between the Earth and outer space. As you climb in altitude, the atmosphere gets less dense as you go. You need to get about 350 km (220 miles) above the Earth's surface to get to the point where a spacecraft can orbit the planet. Any lower than that, and the spacecraft will bump into too much atmosphere to remain in orbit very long.

Where a free falling surface object size mass falls doubling its velocity each second it continues to fall and its velocity is only able to accelerate from 9.8 meters per second to 19.6 meters per second and so on – as time relative to the free fall continues. In the same instance, we discover that in space a spacecraft traveling light velocity without the presence of any type of regular gravity unity like earth's 9.8 meters – in space traveling light speed actually means traveling at 372,000 miles a second. Why is this? It is based on the lack of gravity or lack of propulsion that occurs relative to Einstein speed constant that limits light velocity arithmetic. In space where as a vacuum space measures zero point gravity or $E=mv^2-g^2$ (v^2) the equation can be viewed without the affection of mass $Ep=v^2/-g^2$ without any limitations. Considering the big bang event happened and is greater than the light constant value – within the space grid we can safely assume velocities faster than light are possible especially when we intervene with the lack of any gravity presence in space. The pressure bending space so to speak – illustrates a greater energy (propulsion) value than the light constant as well by itself assuming galaxy's present and impressing on the fabric also are to be included. Planetary galactic alike matter curves and bends space relative to the universe's orbitration and or speed assumed relative to light velocity pushing against the dark fabric of space. It does not rip through but only is allowed to impress against it. Since lacking the quantity of gravity in a vacuum space allows acceleration to increase based on the capacity of propulsion even at light speed measures more in the mathematics deciphered in Quanta Physics Theory and 372,000 miles a second speed and difference to the light speed limit.

Where planets curve space a spacecraft traveling faster than light can only curve and twist the fabric as well especially comparing its size and weight with a planet, star or galaxy. The structure of the impression does not change just because it is a ship traveling through space and or the deeper fabric of it. Creating a wormhole bending and twisting through the fabric only short runs the distance traveling atop the fabric where galaxies, planets and stars form a definite fabric curvature between the deities of space. A wormhole is just that a short cut through the fabric traveling through un-seeable terrain within passing through the fabric from a surface towards a distance planet rather than traveling around atop the fabric surface. Work distilled by Rod Kawecki 2013 has discovered that the possibility of time travel using the wormhole method may not be possible. He has discovered that if the fabric of space which holds gigantic galaxies amidst space leaning against the fabric of space than how can a spacecraft rip through it. A wormhole allows traveling from point 'a' to point 'b' opening up at the destination nearby some planet on the coordinates. But if the fabric of space does not even allow great gigantic planets, stars or even galaxies to rip through how can a spacecraft?

Does Gravity Travel at the Speed of Light?

To begin with, the speed of gravity has not been measured directly in the laboratory--the gravitational interaction is too weak, and such an experiment is beyond present technological capabilities. The "speed of gravity" thought to be a single particle must be deduced from astronomical observations, and the answer depends on what model of gravity one uses to describe those observations.

In the simple Newtonian model, gravity propagates instantaneously: the force exerted by a massive object points directly toward that object's present position. For example, in space even though the Sun is 500 light seconds from the Earth, Newtonian gravity describes a force on Earth directed towards the Sun's position "now," not its position 500 seconds ago. Putting a "light travel delay" (technically called "retardation") into Newtonian gravity would make orbits unstable, leading to predictions that clearly contradict Solar System observations.

In general relativity, on the other hand, gravity propagates at the speed of light; that is, the motion of a massive object creates a distortion in the curvature of space time that moves outward at light speed. This might seem to contradict the Solar System observations described above, but remember that general relativity is conceptually very different from Newtonian gravity, so a direct comparison is not so simple. Strictly speaking, gravity is not a "force" in general relativity, and a description in terms of speed and direction can be tricky. For weak fields, though, one can describe the theory in a sort of Newtonian language. In that case, one finds that the "force" in GR is not quite central--it does not point directly towards the source of the gravitational field--and that it depends on velocity as well as position. The net result is that the effect of propagation delay is almost exactly cancelled, and general relativity very nearly reproduces the Newtonian result.

Kawecki Gravity is altogether different from the two. Kawecki asserts the dominion mass pulling nearby planets and stars towards but is contradicted by the universe's spin pulling everything out towards its edges. The solar system or any system in the universe and or galaxy thereby is impressed in the fabric holding it tight against the fabric it leans into. The solar system orbits and or rotates due to the universe spinning in an orbitration motion at light speed. Light speed in Quanta Physics seems to mean that after the big bang little bang events taken place in the universe - the universe slowed to a specific velocity that seems to be a universal coordinated speed in the vacuum minus zero point gravity. The idea that gravity can be a basis for electricity properties is therefore a part of no the universe geometrics but a state of chemical properties instead. Instead molecular mass is the frequency that holds matter together also known as matter/energy.

This cancellation may seem less strange if one notes that a similar effect occurs in electromagnetism. If a charged particle is moving at a constant velocity, it exerts a force that points toward its present position, not its retarded position, even though electromagnetic interactions certainly move at the speed of light. Here, as in general relativity, subtleties in the nature of the interaction "conspire" to disguise the effect of propagation delay. It should be emphasized that in both electromagnetism and general relativity, this effect is not put in ad hoc but comes out of the equations. Also, the cancellation is nearly exact only for constant velocities. If a charged particle or a gravitating mass suddenly accelerates, the change in the electric or gravitational field propagates outward at the speed of light.

Since this point can be confusing, it's worth exploring a little further, in a slightly more technical manner. Consider two bodies--call them A and B--held in orbit by either electrical or gravitational attraction. As long as the force on A points directly towards B and vice versa, a stable orbit is possible. If the force on A points instead towards the retarded (propagation-time-delayed) position of B, on the other hand, the effect is to add a new component of force in the direction of A's motion, causing instability of the orbit. This instability, in turn, leads to a change in the mechanical angular momentum of the A-B system. But total angular momentum is conserved, so this change can only occur if some of the angular momentum of the A-B system is carried away by electromagnetic or gravitational radiation.

Now, in electrodynamics, a charge moving at a constant velocity does not radiate. (Technically, the lowest order radiation is dipole radiation, which depends on the acceleration.) So, to the extent that A's motion can be approximated as motion at a constant velocity, A cannot lose angular momentum. For the theory to be consistent there must therefore be compensating terms that partially cancel the instability of the orbit caused by retardation. This is exactly what happens; a calculation shows that the force on A points not towards B's retarded position, but towards B's "linearly extrapolated" retarded position. Similarly, in general relativity, a mass moving at a constant acceleration does not radiate (the lowest order radiation is quad-rupole), so for consistency, an even more complete cancellation of the effect of retardation must occur. This is exactly what one finds when one solves the equations of motion in general relativity.

While current observations do not yet provide a direct model-independent measurement of the speed of gravity, a test within the framework of general relativity can be made by observing the binary pulsar PSR 1913+16. The orbit of this binary system is gradually decaying, and this behavior is attributed to the loss of energy due to escaping gravitational radiation. But in any field theory, radiation is intimately related to the finite velocity of field propagation, and the orbital changes due to gravitational radiation can equivalently be viewed as damping caused by the finite propagation speed. (In the discussion above, this damping represents a failure of the "retardation" and "non-central, velocity-dependent" effects to completely cancel.)

The rate of this damping can be computed, and one finds that it depends sensitively on the speed of gravity. The fact that gravitational damping is measured at all is a strong indication that the propagation speed of gravity is not infinite. If the calculation framework of general relativity is accepted, the damping can be used to calculate the speed, and the actual measurement confirms that the speed of gravity is equal to the speed of light to within 1%. (Measurements of at least one other binary pulsar system, PSR B1534+12, confirm this result, although so far with less precision.)

Are there future prospects for a direct measurement of the speed of gravity? One possibility would involve detection of gravitational waves from a supernova. The detection of gravitational radiation in the same time frame as a neutrino burst, followed by a later visual identification of a supernova, would be considered strong experimental evidence for the speed of gravity being equal to the speed of light. However, unless a very nearby supernova occurs soon, it will be some time before gravitational wave detectors are expected to be sensitive enough to perform such a test.

The Grandfather Paradox

(Revised 2012 by this author)

The Grandfather Paradox known by the works of Albert Einstein in 1905 work is the first to suggest time travel into the infinite future. His work also explains the truce to space exploration and the fact that interstellar space seems to be the open entre domain for earth's future. Aside traveling from LA to New York this planet is getting very full of people places and things. For just not that reason space exploration seems to be are only exit. But Einstein's ideas about space travel have some problems that need to be worked out. This is the reason for the paradox. For this reason along with others more technical I will introduce myself. My name is Rodney Kawecki. I am a new author whom believes he holds the fate of future space exploration at his fingertips. Not literally but scientifically.

Trying to turn back Dirac's Sea

Einstein known to be the father of modern physics was a genius but still he didn't know everything and he knew it. But his predictions are noted as scientifically accepted and factual. My work on the other hand is - not publically worldwide known - but should be. The reason is for faster than light speed technology research and involvements to advance physics that should be noted. You see Einstein put a speed limit on universal space flight. My theory on the other hand changes this - and allows for faster than light space travel. The common denominator between my theory and Einstein's is dark matter. Even though Einstein does not directly say dark matter is the underlining element of all material matter and or energies that exist universally. I do. Gravity for instance Einstein claims is a force field - energy type with waves. Electronically waves existing that should be detectable but aren't. But the grandfather paradox he explains fails in Quanta Physics Theory. The reason is stated in this article. Time always flows forwards - never backwards.

If anything there does exist a time paradox in time travel. A time paradox includes information that leads the physicists to believe that while traveling backwards in time is possible - there includes the problem of changing the past. When we review the ethics of time we are also reviewing its paradox simultaneous imagery. What this means is that in today's advance physics we know that light travels at 186,000 miles a second per second. That it is light that is known to be the fastest known particle structure in the Universe. From Albert Einstein's studies he concluded that there exist a time a paradox with it.

The time paradox involves the distance of light having the ability to be at its lengths in two places at the same time. Einstein called this action a simultaneous effect. For the matter of understanding this hypothesis we have to look at light from a perspective that it retains two ends: a beginning and an end. Lights beginning includes the fact that when a light pulse is emitted that a second later and all at once. The pulse reaches its end at almost the same time it starts. This is to say that within a second frame of time - the pulse will exhibit itself simultaneously at two places at the same time. This is faster than anything we can conceive about except in Quantum Mechanics, which measures any infinite basic length to meet with what is called the Planck scale - over a billion billionths of a centimeter.

With this in mind we can look at light as a means to be in two places at the same time in real time. Traveling backwards through the fourth dimension allows for a journey through time based on this analogy predicted by

Einstein in 1905. So the fact remains that if someone had the ability to travel at close to the speed of light. He or she could travel through time - time in the infinite future as Einstein predicted and backwards as well as foretold in Quanta Physics. It is with the mathematics in Quantum mechanics that this analogy can be explained at different the segments of time. Sort of like saying that this time line can be understood more clearly in frames relative to the Planck scale.

Traveling at speeds faster than light have been discovered in real time by Kawecki Quanta Physics that predicts the analog of being able to travel back into the universes past at a time relative to the time yourself existed before the time of the launch of which you traveled back into the past. The paradox of seeing yourself is a real one. But the idea of changing history is not. This is because the time line of which you traveled from is part of the future you used to travel backwards through time into the past. The super lateral paradox by Einstein does not exist. The fact of the matter is that when you traveled from the future you traveled from the future that already existed. Trying to change this fact is impossible. But the idea that if you that's say murder yourself in a place in the planets past will change your own future will not happen, at least not in this life time. The reason is because your future already exist thereby killing yourself in this future event in your own life will only leave the impression that a murder took place and if you were caught you'd go to jail. But by the time any time travel becomes possible - it will far off in the planets future that any type of time travel crimes will take place.

It's based on the idea that Einstein believed time travel to be a loop back in 1905 that he retained his theories on time travel. The fact is that he believed that as a loop - time instantaneously traveled from the 'now' or present strictly backwards without having to travel towards the future to travel into the past as explained in Quanta Physics theory or Q.P.T. The fact remains that he thought that time travel was a two-legged faucet, where a single joint of water is tapped at both ends: one in the future and two the past. He believed time could be reversed physically from the present. Instead of traveling forwards one could travel backwards like one walking on the sidewalk. He thought that since traveling backwards into the past may be possible. That the same person on the sidewalk could actually be walking backwards due to the forwards arrow of time thus traveling backwards in time. As such everything it takes to travel into the future is eliminated in traveling into the past. He designed what are called wormholes and traveling through distorted space as a means to make up for the difference. But the fact is that time travel is conducted by traveling with the arrow of time forwards into the future and increasing to faster than light space velocities this allows traveling faster than the rotation of the planets.

Truth or Fact:

Traveling faster than a planet's rotation to achieve a velocity sounds quite incredible. But the theory in mathematics coincides. These types of facts are what theory is developed from. The speed of light is quite fast also. But the facts render this possibility as an effort to visit other worlds aboard open space. Light speed is quite slow when we research the facts that render the total possibilities of space flight. In my third book I will explain this possibility. The light speed travel limit created by Albert Einstein in relativity fails in Quanta Physics Study. The facts render the truth to evidence for the theory. If we look at the whole big picture like string theory has tried to open up to us - we'll discover a collision of universes in multiple terms of definition is really a big Universe equals like ten of them making a single Universe where we live today. This is fantastic wouldn't you say.

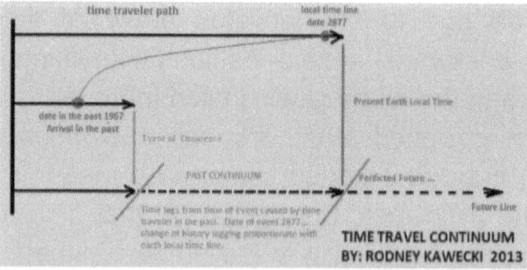

TIME TRAVEL CONTINUUM
BY: RODNEY KAWECKI 2013

Time actually in theory lags behind from the time of the event the future time traveler causes in the past. The future doesn't change except after the date of the past event the traveler journeyed to change history from. The local time frame and predictable futuristic time line continues.

Lacking of grounds for connection chemistry for the paradox

The earth's surface motion activity in the planet's atmosphere density and element particles the planet rotates at 18.5 miles a second...measuring this velocity with the light speed limit 186,000 miles a second earth gravity acts as a weak force turning at only 9.8 meters a second in the spin. Using the earlier planetary velocity chart gravitons pressed against the fifth element vacuum space:

186.000 m/s

-18.5 m/s

Equal: 1,000 gravity tons

We can consider these equations with the planet at rest not including the universe's angular spin attraction towards the outer rim and or the galactic attraction of the solar system dragged towards the galactic outer rim. These actions contradict the dominion sun's attraction towards its gravity well. It is this action that the planets and stars in systems are aligned from being collapsed towards a leading dominion star like our sun throughout the universe. The idea that a 1,000 graviton's the energy of a thousand earth' planets and gravity ton's per square inch mathematically measures equal to the proportionate mass of one of our suns the mass of our sun eq. 100,000 g tons measures the amount of energy mass it would take to penetrate and meet the criteria to time travel into the earths past. The problem that there exist no connection between the earth physical motion and our space ship retaining the capacity propulsion value to make such a journey there exist no physical connection that could account for the physical chain reaction or change parallel activity to make it happen except empty space.

The only probable possibility for time travel in today's science exists on the theory that a portal wormhole inside the planets motional atmosphere could a physical connection be made. The galactic universe matter drags empty space fabric with it as it spins angular momentum. In turn – the galaxies pull and circular planets and stars inside its domain. Inside the solar systems that exist more orbital activity takes place amongst the intermediate planets like what happening in our solar system today. This type of connection is what creates a time arrow coordinates we measure time with using the physical rotations and orbits of our planets. Cutting to the chase – we see that a space ship trying to change time venturing a time travel journey sits dormant in empty space without any real physical connection with the planets around it dominion. The effect our sun has on our nine planets is caught in a physical cycle it retains with all the planets in its domain. An existing space craft retains no physical connection like that. It resides local adrift the cosmos. For time to physically change the space ship would have to be powerful enough to slow drag the existing and connecting planetary spheres and control their orbit in exchange. Make them rotate backwards by pulling the whole solar system into a reverse attraction spin making them orbit and rotate in the opposing direction they are universally set by. For a ship even having he capacity to travel light speed exist no real possibility to this type of space maneuverability.

Time travel on earth even inside the earth's motion atmosphere chambers would have to have the capacity to slowly reverse the physical nature of earth elements and activity. The primary activity would have to be at a global scale and not intermediate to a single room or lab or sort. To understand the concept to time travel more deeply observe using a passage wormhole as relativity asserts. The physical nature of the fabric itself illustrates that not even a star or planet, galaxy or nebula can pen titrate through the fabric grid. And in order to time travel this activity would have to apply. On a flat surface like our earth ocean passing above this grid is our space craft passing through a wormhole means passing into the ocean liquid itself through a developed portal or the ship pulling through the water under the surface grid. It travels and surfaces somewhere else a distance away on the surface there but is a little more complex than that explanation. Our ship travels into the wormhole passing through the portal towards the direction of its destination planet. But like the planets, stars and even the weight of galaxies don't penetrate the surface grid but only maneuver in orbit and rotation pulled by the angular spin they developed in the first place by the weight of their mass or masses. The universe generating an angular orbit slow drag of cosmos matter they do not penetrated the grid surface. The logic of this explanation alerts us as space explorers that specifically the surface grid cannot be penetrate at any polarity or speed as shown to us by the big bang theory itself. A creating occurrence that spun atop a surface grid impressing its presence on top the grip surface forming elliptical dents which by physical terminology and theory penetration can only occur through the physical rebirth of cosmos entity celestial type pulling through the grid developing a physical flat surface type on it to exist. Wormhole technology is still theoretically inconclusive to date.

CHAPTER FIVE

GRAVITY IS NOT AT ALL ENERGY

Gravity decreases with altitude, since greater altitude means greater distance from the Earth's center. At the Interstellar stage it ceases to exist in the same form as a detectable prorogating wave.

Two stars of similar mass are in circular orbits about their center of mass. *(energy acting as a similar 'effect' and *rolling* due to their shapes)

Two stars of dissimilar mass are in circular orbits. Each will rotate about their common center of mass in a circle with the larger, the dominion mass having a smaller orbit. *(object masses are measured outside the sphere, whereas, ' gravitation ' is the actual effect of objects bulk energy mass, acting adjoins with its mass towards the smaller mass attracting it to a specific distance). Gravity is commonly confused with the magnetic force, which actually retains the balance of an objects position and not the distance between objects.

Gravitation is the force of attraction that exists ""not-between"" all particles with mass in the universe. G-force refers to either the force of gravity on a particular celestial body or the force of acceleration anywhere. G-force is measured in g's, where 1 g equals the force of gravity at the Earth's surface (9.8 meters per second per second). As Einstein realized, the force of gravity and the forces of acceleration are mutually indistinguishable on the subject; a person in an opaque box experiencing a g-force would be unable to tell whether its origin lies in acceleration through space or a gravitational field unless they had some way of peeking outside the box. Analysis of g-forces are important in a variety of scientific and engineering fields, especially planetary science, astrophysics, rocket science, and the engineering of various machines such as fighter jets, race cars, and large engines.

Interstellar Gravity Distances

Le Sage's theory of gravitation is the most common name for the kinetic theory of gravity originally proposed by Nicolas Fatio de Duillier in 1690 and later by Georges-Louis Le Sage in 1748. The theory proposed a mechanical explanation for Newton's gravitational force in terms of streams of tiny unseen particles. Georges-Louis Le Sage (June 13, 1724 in Geneva, † November 9, 1803 in Geneva) was a physicist and is most known for his theory of gravitation, Nicolas Fatio de Duillier (alternative names are Facio or Faccio) (26 February 1664 – 12 May 1753) was a Swiss mathematician known for his work on the zodiacal light problem, for his very close relationship with Isaac Newton, for his role in the Newton v. Leibniz calculus controversy, and for originating the "push" or "shadow" theory of gravitation.

Astronomical distances are often measured in the time it would take a beam of light to travel between two points (see light-year). Light in a vacuum travels approximately 300,000 kilometers per second or 186,000 miles per second.

The distance from Earth to the Moon is 1.3 light-seconds. With current spacecraft propulsion technologies, a trip to the moon will typically take about three days. That means light travels approximately two hundred thousand times faster than current spacecraft propulsion technologies. The distance from Earth to other planets in the solar system ranges from three light-minutes to about four light-hours. Depending on the planet and its alignment to Earth, for a typical unmanned spacecraft these trips will take from a few months to a little over a decade.

Intergalactic travel involves distances about million-fold greater than interstellar distances, making it radically more difficult than even interstellar travel.

Gravity is the persistence of planetary and/or surface items mass energy having at a zero point field density volume acting equivalently to attain what is called 'mass-weight". In reference to Interstellar space and intergalactic space fields gravity is based on the vacuum density measured in negative volume causing all materialistic matter to be weightless as a singularity mass entity.

In 1667, Newton mathematically deduced the nature of gravity, demonstrating that the same force that pulls an apple down to earth also keeps the moon in its orbit and accounts for the revolutions of the planets.

Newton openly stated that he had no idea what gravity actually was. All he knew was that it had to be caused by something. Quanta Physics expresses the fact that gravity is defined as an atomic force of all matter created by the big bang.

Objects near the earth experience an accelerating force. The force averages 32 feet per second, per second. Centuries of measurement have firmly fixed the value of G at $6.673 \times 10-11$ cubic meters per kilogram per square second. This was measured and accounts for the earth's surface energy 'effect' known as the 'gravity force'. But is adjoined by vacuum gravity that measured at zero point field energy or $10 -33cm^3$. On earth gravity is reconciled with the planet's surface atmospheric energy amount. If G varies under any circumstances, scientists would have to rewrite virtually every physical law and a long–accepted feature of the Universe isotropy, or the condition that a body's physical properties are independent of its orientation in space.

At the atomic level, gravity is minuscule. In the atom, gravity is 40 orders of magnitude smaller than the electrical force. It is impossible to say under this circumstance if gravity exists as a separate entity at the atomic level, or if it is a resulting, innate part caused by charge separation.

Classically, gravity is always associated with mass. The subatomic particle has a measure of mass, although it is measured using its own electric and magnetic properties. The gravity factor is assumed to exist with the charges and their mass. If it is a constituent of every atom, we can hypothesize that every atom has instant communication with every other atom. The communication would be through a dimension in the atom that we can only recognize as, and call, "gravity."

It seems that "gravity" may be a legitimate dimension in the same sense as we consider length, time, mass, and charge as dimensions. "Gravity" is at the center of every atomic mass and thereby exists only in association with electric and magnetic forces as Quanta Physics interposes it to be. If gravity is one of instant information transfer in another dimension, its relationship to electric charge in our 3-dimensional space-time world is going to be much different than earth's gravity field.

Since the gravity of space is expanding at the Hubble Constant, but is traveling at the speed of light, than thoughts of the universe pictured as an evolving disk, attains the universe retains a centralizing core center or vortex keeping it in motion.

The idea that ...the universe having too little mass -- if omega equals less than one -- than the universe would expand forever, growing ever more tenuous. If omega equals more than one, then the universe would collapse of its own weight, contracting in what is called the Big Crunch.

Is Gravity Push and Pull that gives balance to the universe?

Gravity it is not all energy yet is a fundamental underlying force in the universe. The amount of gravity that something possesses is proportional to its mass and distance between it and another object. This relationship was first published by Sir Isaac Newton. His law of universal gravitation says that the force (F) of gravitational attraction between two objects with Mass1 and Mass2 at distance D is: F = G(mass1*mass2)/D squared.

(G is the gravitational constant, which has the same value throughout our universe.)

When objects are in orbit around each other, there is a strong pull of gravity between them. For example, we commonly say that the Moon is in orbit around the Earth. However, the Moon pulls back on the *Earth elliptical swell*. This changes the Earth a little. One way we see this happening is the ocean tides.

Gravity neither is nor made of energy or any parallel substance related to the elements we know today in science. As all todays elements act similar to one another atomic chemistry the *dark element substance or the fabric gravity space* is un-similar to our elements in physic. As the earth's moon orbits above the planet its elliptical path forms stress or tension on the earths elliptical path creating changes on the earth's surface. Where celestials' lay impressed in the fabric gravity – elliptical paths form gigantic dark element matter chumps around the sphere rotating in an orbit. The moon for instance stretches and pulls on the earths elliptical fabric causing the planets oceans high and low tides depending on the distance the moon shades orbiting the earth during the month of the calendar.

When we review this fact about gravity we discover that it is the fabric that curves and bends adding hills and mounds around celestials like earth. These hills and mounds are what create a specific effect on other celestials and cause say the tides on earth cause by the moon's orbit above it. It bends the fabric close to earth causing the earth to bend sideways a little more to the way of the moon and causes higher tides in the earth's ocean. We have high tide when the moon is orbiting on one side of the planet and low tides when it is further away from it. (It is not all energy)

On Earth

Mass is important because of two major factors affecting how things move in space: inertia and gravity. The more mass something has the more of both its experiences. That is why heavy things (things with a lot of mass) are hard to move. When an object is sitting still, it resists moving, and the more mass it has the more it resists. The amount of thrust needed to move something and how fast it ends up moving are both directly tied to its mass. On the other hand, once something massive starts moving, it is very hard to stop. This is also due to the relationship between mass and inertia.

Gravity is also proportional to how much mass each thing has. The bigger an object is, the larger the gravitational pull it exerts.

Because of gravity and inertia, the more massive something is, the harder it is to get into space, the harder it is to keep it there, and the harder it is to move it where you want it to go when it is there. For that reason, one of the goals of the New Millennium project is to make lightweight spacecraft.

Gravity is a very important force. Every object in space exerts a gravitational pull on every other, and so gravity influences the paths taken by everything traveling through space. It is the glue that holds together entire galaxies. It keeps planets in orbit. It makes it possible to use human-made satellites and to go to and return from the Moon. It makes planets habitable by trapping gasses and liquids in an atmosphere. It can also cause life-destroying asteroids to crash into planets.

We can think of gravity as the effect of the universe in angular momentum causing the planets and galaxy's to weigh into motion curving and creating elliptical impressions throughout space. These elliptical paths and curves in space are what create different effects on a nearby celestial's position and path. as the shape of an open orbit (as distinct from a closed elliptical orbit), such as the orbit of a spacecraft during a gravity assisted swing-by of a planet or more generally any spacecraft exceeding the escape velocity of the nearest planet, as the path of a single-apparition comet (one travelling too fast to ever return to the solar system), as the scattering trajectory of a subatomic particle (acted on by repulsive instead of attractive forces but the principle is the same.

As the shape of an open orbit (as distinct from a closed elliptical orbit), such as the orbit of a spacecraft during a gravity assisted swing-by of a planet or more generally any spacecraft exceeding the escape velocity of the nearest planet, as the path of a single-apparition comet (one travelling too fast to ever return to the solar system), as the scattering trajectory of a subatomic particle (acted on by repulsive instead of attractive forces but the principle is the same.

It is my belief that gravity does not act as today's physicists believe it does and attracts by energy or acts by pulling us. I think push is a more accurate concept it acts as resistance between material deities' both celestial and smaller planetary surface atmosphere dominion. Resistance seems to be the subtle wording when I speak in Quanta Physics Theory science. A heavenly body such as the earth displaces the fabric of space-time, and the result is an effect of an applied tension against the surface of the planet. Matter is pushed down in an elastic-like effect. The tautness of curved space-time holds us down. The fabric of space-time, which covers the earth, is curved because the massive presence of the planet earth distorts the fabric. The fabric of space-time is invisible matter, and although we cannot see the fabric of space-time we can observe its effects.

Naturally, the gravity would be greater on a planet with greater radius and mass than the earth because the fabric surrounding the planet would be stretched to a greater degree of tension. Objects on these planets are "pushed down" with the greater force. The greater force is the activity of the universe as a whole. The tension the universe as a whole puts on the celestial deity's like the galaxies and planets and larger suns are pressed greatly against the fabric of the space field and acts to hold these deities within position. The universe orbits steady at light speed where the galactic bodies are pressed into the fabric skin. We should assume that the lesser deities the galaxy's and so forth orbit at a much slower rate than the universe itself. One of the effects of this is the tension pressed on them by the universe as it rotates and spins. ^The galaxies are slowed down to a specific velocity or spin set inside a fabric hole their impressed into by the universe's spin. They are set in motion as galactic matter pushes its way into its angular orbitration itself pressing itself into the space fabric. As a whole the force of all the matter existing in the universe weighs the pushing force the universe as a whole rotates at. The universe is et in terminal velocity made as it has reached the greatest velocity it can orbit at measured by its weight and impression against the fabric. So terminal velocity not only defines the free fall of planet surface velocity but also the determined velocity of the universe also.

Physicists over the decades have always wondered why the speed limit seems to be ghost like researching the characteristics of its poverty's. But like everything in the universe there always seems to correspond into a common deciphered mathematical conclusion. *Terminal velocity* is one of the common mathematical puzzles. Any object or mass engaged in an immeasurable velocity drive will reach terminal velocity when it reaches the velocity parallel with its maximum weight velocity properties. It can increase velocity no more unless by some external added force placed to it. But by itself with no type of internal or external gravity friction applied as it follows a path – it will not and this includes the universe as well as spacecraft's or space vessels traveling in the Cosmo' gain any more speed. If physicists ask the question why the speed constant parallel to the light zero mass quantities exist we should say that it caused by the universe' terminal speed quantity.

Planetary celestial objects like planets and stars as well as galaxies all form elliptical passages impressions I the dark fabric space field element. It is the planet accelerated impressed weight that these planet elliptical; paths retain formation. They do not diminish as fast as they are formed. As the planet orbitration or elliptical path continues their path positioning may vary but not too much so they are act mostly at a specific continuity. The universe as a whole acts infinitely time wise existing in the cosmos. It would rarely change to any degree physically except by some gigantic force physically applied outside its region.

Elliptical paths are highways that exist created by then presence of planetary galactic activity throughout space. They act as paved roads dividing sections between other existing planetary systems near them.

The terminal velocity of a falling object is the velocity of the object when the sum of the drag force (Fd) and buoyancy equals the downward force of gravity (FG) acting on the object. Since the net force on the object is zero, the object has the speed of object increases, the drag force acting on the object, resultant of the substance (e.g., air or water) it is passing through, increases gravity is not friction that slows an automobile when it ceases accelerating the gas pedal. Friction occurs when the force applied ceases to exist – the car slows to a stop. Energy is a specific mechanism for turbine farce maneuvering. It has nothing to do with the gravity force. Energy or friction occurs again by the lack of propulsion or push. It acts as a centered mass quantity that quantization creates resistance between other objects masses (not attraction) for this reason gravity actually is a weak resistance between object masses. At some speed, the drag or force of resistance will equal the gravitational pull on the object buoyancy is considered. At this point the object ceases to accelerate and continues falling at a constant speed called terminal velocity (also called settling velocity). An object moving downward with greater than terminal velocity (for example because it was thrown downwards or it fell from a thinner part of the atmosphere or it changed shape) will slow down until it reaches terminal velocity. Drag depends on the projected area, and this is why objects with a large projected area relative to mass, such as parachutes, have a lower terminal velocity than objects with a small projected area relative to mass, such as bullets as zero acceleration.

The downward force of gravity (Fg) equals the restraining force of drag (Fd). The net force on the object is then *zero*, a measurable degree of weightlessness and the result is that the velocity of the object remains constant.

Gravity decreases with altitude, since greater altitude means greater distance from the Earth's center. At the Interstellar stage it ceases to exist in the same form as a detectable prorogating wave.

Two stars of similar mass are in circular orbits about their center of mass. *(energy acting as a similar 'effect' and rolling due to their shapes)

Two stars of dissimilar mass are in circular orbits. Each will rotate about their common center of mass in a circle with the larger, the dominion mass having a smaller orbit. *(object masses are measured outside the sphere, whereas, ' gravitation ' is the actual effect of objects bulk energy mass, acting adjoins with its mass towards the smaller mass attracting it to a specific distance). Gravity is commonly confused with the magnetic force, which actually retains the balance of an objects position and not the distance between objects.

Does Gravity Travel at the Speed of Light in space?

To begin with, the speed of gravity has not been measured directly in the laboratory--the gravitational interaction is too weak, and such an experiment is beyond present technological capabilities. The "speed of gravity" must therefore be deduced from astronomical observations, and the answer depends on what model of gravity one uses to describe those observations.

In the simple Newtonian model, gravity propagates instantaneously: the force exerted by a massive object points directly toward that object's present position. For example, even though the Sun is 500 light seconds from the Earth, Newtonian gravity describes a force on Earth directed towards the Sun's position "now," not its position 500 seconds ago. Putting a "light travel delay" (technically called "retardation") into Newtonian gravity would make orbits unstable, leading to predictions that clearly contradict Solar System observations.

In general relativity, on the other hand, gravity propagates at the speed of light; that is, the motion of a massive object creates a distortion in the curvature of space-time that moves outward at light speed. This might seem to contradict the Solar System observations described above, but remember that general relativity is conceptually very different from Newtonian gravity, so a direct comparison is not so simple. Strictly speaking, gravity is not a "force" in general relativity, and a description in terms of speed and direction can be tricky. For weak fields, though, one can describe the theory in a sort of Newtonian language. In that case, one finds that the "force" in GR is not quite central--it does not point directly towards the source of the gravitational field--and that it depends on velocity as well as position. The net result is that the effect of propagation delay is almost exactly cancelled, and general relativity very nearly reproduces the Newtonian result.

This cancellation may seem less strange if one notes that a similar effect occurs in electromagnetism. If a charged particle is moving at a constant velocity, it exerts a force that points toward its present position, not its retarded position, even though electromagnetic interactions certainly move at the speed of light. Here, as in general relativity, subtleties in the nature of the interaction "conspire" to disguise the effect of propagation delay. It should be emphasized that in both electromagnetism and general relativity, this effect is not put in ad hoc but comes out of the equations. Also, the cancellation is nearly exact only for constant velocities. If a charged particle or a gravitating mass suddenly accelerates, the change in the electric or gravitational field propagates outward at the speed of light.

Since this point can be confusing, it's worth exploring a little further, in a slightly more technical manner. Consider two bodies--call them A and B--held in orbit by either electrical or gravitational attraction. As long as the force on A points directly towards B and vice versa, a stable orbit is possible. If the force on A points instead towards the retarded (propagation-time-delayed) position of B, on the other hand, the effect is to add a new component of force in the direction of A's motion, causing instability of the orbit. This instability, in turn, leads to a change in the mechanical angular momentum of the A-B system. But total angular momentum is conserved, so this change can only occur if some of the angular momentum of the A-B system is carried away by electromagnetic or gravitational radiation.

Now, in electrodynamics, a charge moving at a constant velocity does not radiate. (Technically, the lowest order radiation is dipole radiation, which depends on the acceleration.) So, to the extent that A's motion can be approximated as motion at a constant velocity, A cannot lose angular momentum. For the theory to be consistent there must therefore be compensating terms that partially cancel the instability of the orbit caused by retardation. This is exactly what happens; a calculation shows that the force on A points not towards B's retarded position, but towards B's "linearly extrapolated" retarded position. Similarly, in general relativity, a mass moving at a constant acceleration does not radiate (the lowest order radiation is quadruple), so for consistency, an even more complete cancellation of the effect of retardation must occur. This is exactly what one finds when one solves the equations of motion in general relativity.

While current observations do not yet provide a direct model-independent measurement of the speed of gravity, a test within the framework of general relativity can be made by observing the binary pulsar PSR 1913+16. The orbit of this binary system is gradually decaying, and this behavior is attributed to the loss of energy due to escaping gravitational radiation. But in any field theory, radiation is intimately related to the finite velocity of field propagation, and the orbital changes due to gravitational radiation can equivalently be viewed as damping caused by the finite propagation speed. (In the discussion above, this damping represents a failure of the "retardation" and "non-central, velocity-dependent" effects to completely cancel.)

The rate of this damping can be computed, and one finds that it depends sensitively on the speed of gravity. The fact that gravitational damping is measured at all is a strong indication that the propagation speed of gravity is not infinite. If the calculation framework of general relativity is accepted, the damping can be used to calculate the speed, and the actual measurement confirms that the speed of gravity is equal to the speed of light to within 1%. (Measurements of at least one other binary pulsar system, PSR B1534+12, confirm this result, although so far with less precision.)

Are there future prospects for a direct measurement of the speed of gravity? One possibility would involve detection of gravitational waves from a supernova. The detection of gravitational radiation in the same time frame as a neutrino burst, followed by a later visual identification of a supernova, would be considered strong experimental evidence for the speed of gravity being equal to the speed of light. However, unless a very nearby supernova occurs soon, it will be some time before gravitational wave detectors are expected to be sensitive enough to perform such a test.

The key words in trying to determine whether gravity travels at the same velocity as light is weighed by the fact that the universe has reached a specific point in evolution that all matter moving in weightless motion has reached a light speed velocity. To answer the question why light speed seems to by the common denominator in the cosmic question relays on the theoretical idea that after the big bang that formed the universe its accelerated velocity of the explosion had to have been much greater than massless quanta velocity chemistry. The mass size of the cosmic egg had to exist and be so gigantic to an extreme size that everything that exist in it today the force had to have been infinite in nature. To wonder what the measurement of infinite is from this point of view is beyond common sense. We could go on to say that the big bang is the theoretical conscious of egotistic imagination that allows humans to achieve notoriety and common sense about life.

The big bang event scattered divine planet matter throughout the cosmos in every type of form and chemistry that exist and is still unknown to mankind. The dominion substance we call "The Fabric of Space" built elliptical wave paths and caused matter from the bang to immobilize and curve in orbital celestial rounded spheres. Brighten stars and scattered gaseous nebulous. Celestial's spheres and galactic embryus cosmos at its earliest time evolved at a rapid rate forming into different condensed chemistry deity's we see as today in the skies. Trillions of eons pasted as these cosmic eggs slowly matured. At the greatest point these embryus all arrived from an unknown darkened shadow of the unknown behind the fabric space matrix surface. That area of the dark elements that hides itself from energetic light quanta' rays. Behind the indivisibility of what we observe as Dirac's Sea lays the matrix construction of dark invisible blocking blocks existing of the chemistry of structural matter of celestial types and invisible life existing creatures trapped behind the crest lined grid that divides our world from the builders.

If you've ever looked at a small creature captured from the furthest depths of the ocean floors whose creatures that have ever even seen light – or the reflection of it lay at the sea bottom as invisible living creatures that if captured by a n ocean craft topside at the sea surface, a captured underwater cockroach published by the discovery history channel in 2006 captured by accident in a clear glass it was only by the reflection of sunlight that this invisible creature became noticeable by the crew personal. Does this prove that life may have formed from the darkest depths of our ocean floor? Not quite.

But what it does indicate is that there exist the divine possibilities that in space and in the depths of the unknown most likely and to a specific degree of possibly probability noting that the celestial space field grid retains a specific surface grid that this grid or surface shield may act as a division between how life evolves from inside the darken shadows and deepest depths of Dirac's Sea in space and somehow surface life type material planetary quantities in celestial form that evolve outside the hidden grid depths.

A lot of physicists believe that starting from the big bang event and primeval point that all came to exist from nothingness. Early royal religious societies believed the in the divine nature of a living god that all that exist came from the father god. And that mankind had no say so to argue that degreed. Some scientist and physicists in the earliest times were confined for acts of treason for acting against this creed. Family members killed as being witches and Isaac Newton for one was placed under house arrest for the rest of his life. The idea that science separated the creed about aspects of researching technologies about the universe went against our load were very much punishable at that era.

Reviewing back to the facts about the big bang event or even smaller little bangs that formed cosmic eggs formed through embryus eggs at interior dense positions inside the big bang egg that scattered formed later into gigantic galaxies, Nebulas etc., throughout the cosmos. Having the embedded dense chemistry having existed inside the larger cosmic egg exploded later into a wide spread of existing galaxies. Inside them smaller explosions forming gigantic suns and molding planets and divinely bright stars creating their own existing disk plate formations like are solar system and the division of systems around it. The universe was not at all created at a single momentum in time but scattered in time frames scattered throughout cosmic terminology over eons.

Chapter Six

Wine and Roses from Venus

The Universe is commonly defined as the totality of existence including planets, stars, and galaxies, the contents of intergalactic space, and all matter and energy. Similar terms include the cosmos, the world and nature.

The observable universe is about 46 billion light years in radius. Scientific observation of the Universe has led to inferences of its earlier stages. These observations suggest that the Universe has been governed by the same physical laws and constants throughout most of its extent and history. The Big Bang theory is the prevailing cosmological model that describes the early development of the Universe, which is calculated to have begun 13.798 ± 0.037 billion years ago. Observations have shown that the Universe appears to be expanding at an accelerating rate.

There are many competing theories about the ultimate fate of the Universe. Physicists remain unsure about what, if anything preceded the Big Bang. Many refuse to speculate, doubting that any information from any such prior state could ever be accessible. There are various multiverse hypotheses, in which physicists have suggested that the Universe might be one among many universes that likewise exist.

Instantaneous light speed theory is a speed coordinates theory for space aviation travel. It relies on the idea that time dilation, time travel and traveling to slow a clock mechanism is unrealistic theory. According to Albert Einstein all the above mentioned ideas are a fact of his special relativity and general relativity both developed in the earliest twentieth century. New theory about space travel has now been excited into a new era of technology. One of those technologies is instantaneous light speed theory.

This new light speed based theory was designed by Rodney Kawecki also the author of numeral books on relativity theory that he has reinvented into argumentative supplementary. Quantum light speed instantaneous theory refers to the idea that a space cannot hold and drag a multiple planetary orbitration cycle of planets to cause the planetary cycle to change course or the arrow of time it itself coordinates. To look at it more realistically we can view a mosquito many times smaller than what it measures on earth place it next to a planet or multiple planet cycle of a solar system and believe that the mosquitoes speed will cause the said planet time to maneuver backwards or forwards. To look at his realistically it can't be done. Time travel is thus thrown out the window. Maybe and possible modern might be missing some great equation in space theory and mechanics but as it stands now - time travel is not possible according to this new insight.

Time travel, time dilation do not occur without an effect acting on it as gravity does with an external force. In space there exist no external force except possible the grid itself but Rodney Kawecki has also researched the design of supernova experience and methods where he concludes that a vacuum density not really knowing how deep it can maneuver even by the polarity of the big bang that also retains an effect resistance response accordingly that a StarCraft cannot maneuver and drag the weight of vacuum density to lure and change the cycle of planets. No according to Kawecki it seems' all so impossible.

What Kawecki has resign to is the fact that even though time can coordinate backwards towards the number zero as velocity increases - 'instantaneous light coordinates velocity or $v2/1,2$, illustrates that the fastest a starship can maneuver to is zero to $v2/1,2,3,4$ etc...traveling from point 'A' to point 'B' is the shortest length of time a starship can possibly travel. Time cannot run backwards or faster than it was designed for except by mechanical manipulation of change of space. Realistically time does not run backwards. It is used to measure the arrow of our universe's maneuvering cycles that all travel forwards. And no wormhole exist or can be modified to drag gigantic planets into its web slow drag even if it could bring a planet into it. Which it can't and that's about what it would take. No instantaneous light velocity theory stands as the fastest means for potential velocity to conquer distance and time. Velocity does not travel backwards though some physicists think it might. Time shortens the distance by using velocity but time only peeks to zero point units.

Quanta Physics Study is a newly developed theory on advance physics which advance the theory of Relativity design by Albert Einstein in 1905. Relativity asserts that nothing can travel faster than light. This assumption made by Einstein takes for granite that 'light' as an element having no mass resolves the universal speed limit. But it doesn't. It doesn't because a theory which advertises FTL...assumes the characteristics of a new particle other than the 'quanta'. Quanta Physics developed by Rodney Kawecki of Los Angeles California has designed a universe that does not limit universal travel to the speed of light. How is that? You might ask.

QUANTA' PHYSICS THEORY

The idea in Quanta Physics Theory is that by doing the math faster Than Light theory intervenes with SRT and GRT and also Quantum Dynamics by Max Planck 1925. That such a flight travel of FTL does not include a massless particle but that which involves weight relative to speed. As such 'light' becomes exempt in the dialogue. The theory to FTL mathematics include the intervention of 'e' energy and its potential source relative to the object in travel. Included with special relativity and Einstein's equation concerning the speed of light. Under the same circumstances. FTL is possible measured with the fact of material weight and empty weightless space. Relativity is a theory asserting gravitation. Empty space...has no gravity. The theory of special relativity and the light constant become exempt in the equation for FTL because of this. The weight complexity - becomes balance in the flight pattern at light speed when its measured with the complexity of empty interstellar space Or the vacuum.

Summed in the mathematical equation the velocity of atomic energy and speed relative with the vacuum the sum equals FTL. (to read more go to specialrelativityquantaphysics.com). The equation is $E=Mg^2T-1$second. e is energy equals m mass q is 372,000 m/s minus one second. In the new theory and equation. The vacuum assumes the density against the mass object at 10-34cm. 10-34cm (empty space) adverts to minus negative density to a positive number which assumes 'one' or the number '1'. The speed of relativity or light asserts the speed of 186,000 m/s. at 10+34cm. The difference is the vacuum figure relative to the positive gravitational planetary sphere. The force of gravity at 9.8 m/s to include this figure as a density we find it to be 10+34cm. The equation therefore asserts that the speed of light in vacuum is measured from 10-34 to the positive number 1 and the acceleration to 186,000 m/s. From 10-34cm we then have 186,000 m/s. From zero or one to 10+34cm we have another 186,000 m/s. Together the equation sum is 'q' squared at 372,000 m/s. How does this new theory resolve the technological question of space travel in the fourth dimension? Special Relativity opens the door to the infinite future. Quanta Physics opens the back door.

The sum of this new theory and FTL superluminal space travel illustrates a new theory for superluminal FTL doctrine. The facts are real and so is the physical nature relative to the equation with the Universe. Not discovered until 2005 . Quanta Physic is the new developing of the future of space exploration in scientific technology literature. Unlike any the theory in advance physics does Quanta Physics include the addition of its theory to intervene with any other in respect to its mathematical equations and physics description about the Universe. Patented by Trade Mark. Quanta Physics is the newest physics theory that is believed to become the predecessor of SRT,GRT and QED upon its release.

The Kaweckian Cesium Wo

"Einstein's theory of special relativity and the principle of causality imply that the speed of any moving object cannot exceed that of light in a vacuum (c). Nevertheless, there exist various proposals for observing faster-than- c propagation of light pulses, using anomalous dispersion near an absorption line, nonlinear and linear gain lines, or tunneling barriers. Here we use gain-assisted linear anomalous dispersion to demonstrate superluminal light propagation in atomic cesium gas. The observed superluminal light pulse propagation is not at odds with causality, being a direct consequence of classical interference between its different frequency components in an anomalous dispersion region."

Can a light pulse travel faster than the speed of light? This question has intrigued physicists for many years because such an event could violate Einstein's theory of special relativity and the principle of causality (that cause' always precedes 'effect'). Together these imply that no object or information can travel faster than the speed of light, $c=3. 10^8$ m s-1. For nearly two decades, physicists have been sending certain light pulses faster than cover short distances (so-called superluminal propagation), but the light pulses have always been distorted in the process so interpreting these experiments has been difficult.

In May this year, Mugnai reported superluminal behavior in the propagation of microwaves (centimeter wavelengths) over much longer distances (tens of centimeters) at a speed 7% faster than c. A report by Wang of this issue now demonstrates a very large superluminal effect for pulses of visible light, in which a pulse propagates in a specially prepared medium with a negative velocity of -c/310: that is, not only faster than a pulse travelling in a vacuum, but so fast that the peak of the pulse exits the medium before it enters it!

A negative velocity can be understood by comparing the times it would take for identical pulses of light to cover some distance L in a vacuum (travelling at velocity c) and in a superluminal medium (travelling at velocity v). The difference in transit times $.T= L/v-L/c$ is a negative quantity if the velocity is superluminal if v has a negative value then .T can become sufficiently negative that the peak of the pulse emerges from the medium at an instant earlier than when the peak of the pulse enters. This brings to mind Arthur Buller's well-known limerick with relativistic over tones:

But Wang claim that, unlike the heroine of this rhyme, their light pulses do not violate causality. They argue that their superluminal pulses are the result of the wave nature of light itself (fortunately, making it impossible for an object with mass to travel faster than c) and that no actual information, or signal, is transmitted faster than c. They use smooth, well-defined light pulses, so that the peak of the pulse at the output results from the forward rising edge of the input pulse, which occurs far earlier in time, making it consistent with causality. An abrupt feature in the light pulse would not be able to travel faster than c. This means that even if the 'effect' appears to precede the cause', you still can't send useful information — such as news of an impending accident — faster than c.

A light pulse has a finite duration, and it is a well-known theorem in physics (the bandwidth theorem) that, to create a pulse of finite duration, an infinite number of waves of different frequency must be added together. The shorter the desired pulse, the larger the bandwidth of frequencies that must be used. All light pulses are therefore formed by a packet of waves of different frequency, each of which has a different amplitude and phase. There is a distinction between the speed of individual waves, called the phase velocity, vp, and the velocity at which the peak of the wave packet propagates, known as the group velocity, vg. In a vacuum the phase and group velocities are the same, but in a highly absorbing or dispersive medium they are usually different. A negative group velocity results when the phases of the different frequency components are shifted by the medium through which they travel, so that the wave packet they form at the exit is brought forward in time compared with the same pulse travelling through a vacuum.

One way to achieve negative velocity is to modify the refractive index of the medium through which the light passes. Last year scientists at Harvard6 and elsewhere succeeded in modifying the refractive properties of a cloud of ultra-cold atoms to generate very slow light pulses with group velocities of a few meters per second. To create the opposite effect — superluminal pulses of light — you need a medium in which the refractive index changes rapidly, for example near an atomic absorption frequency.. The only problem is that the so-called anomalous dispersion region in where vg can be negative is also in a region where there is increased light absorption. In experiments with such highly absorbing materials, the light pulses are either strongly distorted or absorbed, making any faster-than-light claims difficult to interpret.

Sending photons faster than light

How light absorption and refractive index of a dispersive material change rapidly with wavelength when the wavelength of the light pulse is near an atomic absorption band. The anomalous-dispersion region (where the group velocity of light can be negative) coincides with a region of strong light absorption. How the gain and refractive index of cesium gas changes with wavelength when there is a 'gain doublet' (two closely spaced peaks) in the amplification of light. Wang shows that in this case the anomalous-dispersion region can be used to make a pulse of light travel faster than c.

A more promising approach to making superluminal light pulses is to work with an atomic medium where there is gain (amplification of light waves) at the atomic transition frequency. This is achieved in a laser-type medium by creating a 'population inversion', whereby a higher population of atoms are in the excited than in the lower-energy atomic state. In this case, anomalous dispersion occurs at frequencies lower than the transition frequency. But close to the transition frequency, where the effect is largest, the rapidly changing gradient in the refractive index causes severe pulse distortion. One way round this problem is to use a gain doublet — that is, two closely spaced regions of gain where the zone between has steep anomalous dispersion but without strong pulse distortion. This is what Wang has now achieved.

The experiment by Wang and co-workers creates this type of gain doublet in a six-centimeter cell containing cesium gas by using two laser fields closely spaced in frequency. They first measured the refractive index of the cesium using a third 'probe' laser, and produced a dispersion curve similar to with a steep gradient in the anomalous dispersion region corresponding to an expected vg= - c/310. When they sent a 3.7-microsecond light pulse through the medium, it appeared at the exit of the cell before it arrived at the entrance. Although the pulse itself is only shifted forward in time by a modest fraction (1.7%) of its width, this corresponds to the wave packet leaving the cell 62 nanoseconds before it arrives — in other words, travelling nearly 20 meters away from the cell before the incoming pulse enters it. Compared with the time to travel six centimeters in a vacuum (about 0.2 nanoseconds), the 62-nanosecond lead means that the group velocity of the pulse inside the medium is - c/310, close to the predicted value.

In this experiment, each of the different frequency components making up the pulse experiences a slightly different dispersion in the medium. The relative phases between them are therefore changed and the pulse shape is shifted to bring the pulse wave packet (or group velocity) forward in time. So the anomalous dispersion leads to interference between different frequency components of the pulse that produce the superluminal effect. Although amazing, this type of superluminal pulse propagation does not violate the principle of causality.

There remains, however, some debate about what is the true speed at which information is carried by a light pulse. Traditionally the signal velocity of a light pulse is defined as the speed at which the half peak-intensity point on the rising edge of the waveform travels; in this experiment, this is clearly superluminal. In contrast, some researchers argue that the true speed at which information is carried by a light pulse is not the group velocity of a smooth pulse, but rather the speed at which a sudden step-like feature in the waveform travels, which so far has not been shown to exceed c. Superluminal effects are especially interesting in the case of light pulses consisting of only a few photons, in which it could be argued that the group velocity is the same as the velocity of the individual photons. The type of superluminal behavior discussed here is also predicted to apply to single photons, which might have implications for the transmission of quantum information.

The Interstellar Space Field is a weightless fabric type equilibrium where objects of material are not subject to the same laws that govern a gravities sphere like Earth. Objects in space warp it because of their energy not their weight, thus create a flat terrain inside an impressed pressure of expansion in space.

After reading this journal on faster than light particle impulse boosting at a velocity 300 times the speed of light as explained in this literature. we come to realize that unlike relativity that creates a wormhole in space using an increase of energy to bend the space fabric we can use a gaseous chemical called ' cesium gas". In the experiment the gas is compressed into a cloud chamber where the photon impulse is pushed through to increase its velocity. This experiment has been performed many times with the same results. Another gaseous type cloud chamber used is a 'potassium gas' that emitted the same result but only just beyond the light barrier.

Quanta Physics asserts change in modern physics based on new technologies in the era meaning that over time new experiments that correspond to new technologies and advances are analogue into a new series of events for the advancement of science physics technology.

Is dark gravity a cloud governing all matter

Here Rodney Kawecki has engaged into climaxing a new inventive wormhole possibility into the sciences by using the 'cesium gaseous' cloud chamber for the purpose. The idea here like " The Rosenberg-Einstein Bridge wormhole', reported in 1916, Kawecki has invited a new theorem to invert superluminal progressiveness into the picture by using the 'cesium gaseous' cloud as a means to exhibit and renew faster than light space travel analogy. For all purposes, we find that we don't have to create an erotic space fabric bending machine in space to boost a ship's velocity. It is easy enough to imagine a cloud chamber built in empty space to do the job. And it would be a lot easier than manufacturing a tripod electrical wormhole machine.

The idea is not too hard to imagine. Like today NASA space projects that assemble space stations in outer space are a yearly event in the United States Space Exploration program entailing the possibilities of interstellar space exploration and assembling space station equipment in space. These works in progress today illustrate a new reason why the " Kaweckian Cesium Space Chamber' could not be built instead of using high energy tripod interaction to bend space - if in so the fabric can be actually bent. It just means more convenient that building a Cesium Cloud Chamber in space is more practical and would cost less and it could also be an experiment in space that inserting "matter" through a miniature chamber would telltales a lot of technological information about interstellar space flight as one of its rewards.

"A Massive Dark Cloud Universe"

It is this author's belief that the universe recedes at the center scale of a gigantic infinite in size to be dimensions a dark cloud, which is a part of a much larger universe having all the characteristics as this one. "The Dark Cloud Universe Residence Theory" atmosphere retains negative vacuum atmosphere, is cold with high thermal planetary sun entities that stay within its perimeters as time travel slower than outside the cloud. Outside this dark cloud entity exist stars, galaxies and possibility other universes like this one.

Perceived infinite in nature this description of the "Oscillating Universe" was comprehended by the fact that that which surrounds this universe is blocked by a luminous dark matter material unknown to us in material chemistry. An origin of the cosmos and discover of such cloud like sphere planetary bodies that exist in this infinite realm, the idea that this universe infinite to our measurement and/or nature is actually a small part of a much larger and clear skied universe or terrain of luminous dark matter region. Even though this universe retains shape and resistances due to dark matter outside this dark cloud may exist a cosmic terrain of nature that exceeds this nature and cause for dark matter resistances resulting from futile forces.

According to modern physics and relativity, it is uncertain of what space travel actually consists of. The best explanation is what is called "Alcubierre Warp Drive " which states: a spacecraft travels atop a flat space terrain dark matter the more massive the planet, star or galaxy, the deeper the curvature of space around the body much smaller than a planet or galaxy warps space also relative to its presents on the dark matter terrain. As the ship accelerates the front of the vehicle begins moving against an impressed dark matter below its body. As it gains speed it pushes harder and harder against its warp curvature its body impressed on the space field building up resistances to the applied force of the ship's acceleration. When the ship reaches near light velocity a wave is formed in back of the ship due to its acceleration force on the impressed dark matter fabric, in front of it pressing downwards on the terrain it impressed. The ship now has created a wave-like impression pushed down in front of it and a wave curling behind taking up the slack - which pushes it at the specific velocity close to light speed and it doesn't stop. That achieved velocity due to the absence of exterior (energy/magnetism) interruption impressed like gravity here on earth. No - the ship will continue to travel at that specific velocity and will not stop - this is due to the absence of an exterior energy acting on the spacecraft. (The building blocks of Quanta Physics)

As for 'light' it only travels having no mass atop this dark matter interstellar terrain. The distance between point 'A' and point 'B' are shortened because of its speed. The ship on the other hand having a mass - creates a wave impressed travel continuum that doesn't allow it to slow down at all except by a reverse acting thrust mechanism that would have to be made for this purpose.

A light beam travels the long distance up and around space curvature because it will not travel through dark matter due to its absent mass quantity where a massive spacecraft will light will not except by the point of emit-tense. Here implied is that a dominion planet, sun or star due to its mass size becomes dominion to near planets and are trapping them in its deep curvature warp field.

A ship travels through the dark shadows of space as it pushes against it with its applied force of acceleration. As the ship passes light speed - it creates a closed curvature of space under and behind it of great proportion until finally it ends up surf-riding inside this curved wave. Now - it can travel from point 'A' to point 'B' passing through the dark matter on its own perpetual impressed wave it formed from space itself..

Alcubierre of New Mexico University, believed because the dark matter terrain may be positive energized by nature as Albert Einstein asserted in his theories that creating this space-wave meant forming the Casmir Plates, that form a negative energy resistances bubble inside close to the center of the wave. Quanta Physics shows that this is not the case. That the dark matter is only a density field in nature and retains zero point energy and is in fact a vacuum at 10 - 33cm^3 and because it reads into negative reading as a vacuum it is not an energy source type space. The vacuum created by the universe spins on the universe's axis that is what creates the flat space impression in the field. Because it spins at the speed of light it actually drags along the impressed planets and stars and galaxies along with it but are kept in a still and almost subtle motion that the universe's axis allows the planets, stars and galaxies to actually roll all in the same massive direction.. Dark matter also known as " The Dark Cloud " introduced by Rodney Kawecki, is a futile substance chemistry created by the universe's velocity as it drags massive planetary bodies it also trait-lines information in three dimensions, the pass, present and future era's.

The Dark Cloud "recedes with retained force limitations as well as the lifespan of humans living within it. Possibly outside the Dark Cloud is a more clairvoyant extending universe of which the Dark Cloud can be seen by other civilizations. Alien's observing this Dark Cloud have entered it with their futuristic applications of advance physics more-so than ours and have the capacity to travel great distances in shorter time then us - as we strive more to advance are own technology here inside the cloud. To answer the question of "Why is this universe so dark " is answered but outside may exist a more advanced alien civilization that corresponds with sightings we have had since the dawn of history here on earth.

Interference of this Dark Cloud has retained our civilization to a slow-drug and indirect future with a lot of questions that still have yet to be answered. The reason - this Dark Cloud ' keeping us inside it living at a more slower state than outside it where even more existing planets, stars and galaxies peak the dark realm. Existing inside this vast cloud in a foremost cold dark matter state - the clouds negative densities keeping our universe in check from outside its perimeters limit are advancement in technology from logic. The idea that Time Travel may be possible is caped with the Einstein light constant that otherwise outside this clouded mystery space acts in a totally different manner. As we look at the universe orbiting as a disk we discover that it is outside the disks perimeter that light moves faster than at the middle. A vortex shapes the center created by the big bang event that started this universe and by 'Inflation' due to the explosion the universe has subsided into a slow expansion era from the beginning. Outside in civilizations from other universes accumulated closer to the universe's edge because the universe spin and axis are faster they advanced faster than us. Also and more important we can observe the storing of information of this universe and others by the fact that like ripples in a pond a orbiting universe is catch in the drag of a universe disk spinning at the speed of light but information is saved because the planets, stars and galaxies warp the space in their vicinity. Because of this flexibility of dark matter the planetary bodies throughout this universe exist and grow forwards in the direction of time relative to the universe's structure and developments. Can 'Time' run faster in another universe? I would have to say yes based on the fact that the expansion of existing dark matter that created the big bang in the first place - retains specific negative density that moving ripples across its terrain move along like a river pulling the permanent warp fabric hole in its space with it. Looking at the made material universe we discover that at its edges the disk curls upwards at the top and downwards at the bottom of it curled by the explosion that created bits and pieces of a big bang sphere.

Another question about our universe is why do we as human being of celestial type have to be the "one and only" universe? Can't we be a part of a much larger universe? What we call "our universe" in reality can be only a cluster type dark massive cloud. A cloud surrounding a much greater universe we know nothing if little about. A side from what Stephen Hawkins calls multiple universe theory, we can go further and address the fact that t a dark massive cloud is blocking are observation of a greater and larger universe. The idea that darkness shadows the night sky is it a dimension blocking a more massive universe surrounding us. At the conscious scale we as celestial beings as one might say. when we die are carnal bodies are left behind. Spiritualism tells us the afterlife dimension is that are souls making up the space we are born into in this world becomes a celestial vehicle in a new dimension. When we observe the idea about are universe we observe its structure residing inside a bubble realm. A bubble born after the big bang that takes up space forming the bubble do we see are universe as a bubble of this sort? Or is it just a smaller bubble residing inside a darkened cloud blocking enlightenment of a much larger universe?

The universe as we call it expands at the Hubble Constant which measures about 18 miles per second. For an infinite so to speak universe and taking into consideration its mass this velocity of expansion is very small. Relativity asserted that it is expanding at the speed of light - but it's not. Standard University believes the universe to orbitrate like a flat space type disk even though it is expanding in width at the Hubble Recession it orbitrates at light speed. This analogy has confused a lot of physicists. My own observation of the universe is as Stephen Hawkins claims and that is that the universe resides as a bubble taking up space after the big bang. Like the universe its expansion is very slow in reference to its mass. Much slower if we are blocked from its total view inside this cloud. But like the bubble space theory we can also consider are physical being in the same manner. We take up space. Some call are immediate physical surrounding of our bodies as having a physical aura about it. Is this a same type bubble taking up space like the universe?

When we review these facts about ourselves we discover that at a scientific level are bodies do take up a certain amount of space. It is a part of nature. Physicists also believe that we as human beings evolved in are physical intellect and factional beings with the physical nature and are bodies as well evolving also. But what about are minds and intelligence? We are born and are born into taking up a certain amount of space like the universe. Some scientist believe that are intellect and intelligence also evolve in the same manner but much slower. We are like dust-mites on a planet surrounded by galaxies. As a species do we evolve with the universe? Taking into these considerations I believe we do. Scientist as well observers of science of human development and whom study the universe do say that as a species that we have yet developed to a capacity to answer specific questions about the universe and that until we do develop over time that these questions will be answered in due time. Then again we are confronted about are human intelligence. As dust-mites on a small world inside a much greater universe does are intelligence slower expand like the universe is expanding? Will it take time to solve the problems of this world based on the growth of the universe?

When we confront these types of questions about human nature we discover that like the universe we have come into this world as a species from a compressed state. Like the big bang that formed this universe do are intellect and intelligence develop in the same way? As human beings we were created from the star dust of the earth. We are formed both male and female and we retained intelligence about us. Are minds we product of a celestial manifesting brought out of association between the sexes. As a born species we grow out from within a compressed state somewhat the same way as the universe did. But still we are confronted with are intelligence as human beings. Does are nature make us wait from answering questions arose from are interest about ourselves? Is it a pattern that must be followed where it is only in due time that all will be revealed to us. And what about the dark cloud? Does it's darkness keep us from the real truth or are we not yet intelligent enough to discover these phenomenon's?

As a physicist I observe television as a scope of what is in head for us as it develops over the decades. As a group community and species I think that television directs attention to where we as a species are headed into. It is only that we lack the intellect of what is happening around us that we lack a divine intellect.

When we look at the observed universe we discover that we are not allow but are hidden. Whether in a dark cloud away from the larger universe as mentioned or multiple universe's as Stephen Hawkins predicted. But in all we are not allow in any aspect. We are far away. This aspect of the science I have tried to account for in my writing of faster than light space travel. Unlike Einstein's ideas about space mine are focused on non-gravitational origins. But as a science we have learned that by observing are universe if in fact it really is and not just a large cluster hiding the real realm. We have observed the measurements and changes lengths and widths and have tried to put a census well enough to explain it. But what we do know is that the universe as maintained by the standard theory is that everything that exist in it was once a singularity. Or was it? I have speculated that that like a super nova and light brighter than ever thought it could be hidden from are eye sight the development of planets, stars and galaxies can have developed through independent growth development. That at super infinitesimal deity energy had the capacity to join by attraction and develop small infinitesimal epods as explained in my earlier book that attracted and attracted and grew larger until they formed a deity material fabric and formed into the community of planets inside the galaxy and started to shine. Over time depending on the sizemass of the deity it cooled and a planet was formed in space. Birth-rite of the collective so to speak unlike the Hawkins discovery about possible multiple universe's their development could have formed in the same manner and as the universe or dark cloud we reside in.

Mathematical Proof of the Big Bang in Quanta Physics

According to the standard theory of the universe the big bang occurred with a negative dark matter measurement reading 10-120cm² (Planck units). Quanta Physics asserts increasing this measurement because it was developed by the Einstein light constant - which Quanta Physics shows is not at all - constant throughout the universe at all but is confined by the inertia of a planets realm elements. Quanta Physics has adjusted this reading to add the Interstellar realm of space into its category and its mathematical logic to $E=mc^2$ to 10-240cm². It does this based on the speed limit hypothesis of the light constant that in space can be increased due to the absence of an inertia atmosphere. Again Quanta Physics observes its findings and discovers that an explosion like the big bang induces great mass that related with the dark matter density holding it in place like it does the stars, planets and galaxies by allowing these objects to warp their own position in space. The cosmic egg would have an increasing mass that even though mathematics gives to us specific readings or measurements to these types of causes - it does not calculate endurance and inner mass behavior. The cosmic egg having increased in its mass indifferent to the process we observe it today but by unstable accumulation of infinitesimal particle growth - meaning that before the big bang it is believed by Quanta Physics that unstable infinitesimal energies swarmed the residence of dark matter and assumed upon itself causing independent epods or growth of energy at that time. These energy pods as they are called increased in mass absorbing other nearby energy particles that were infinitesimal in nature and velocity bombarding into the larger epods that exist by absorption through colliding in the more advance mass pods. Thus the cosmic egg was created.

Because of this advancement of infinite mass characteristics in unstable energy these energy pods grew and grew with nothing to stop them by dark matter itself. Enhanced was the ' Dark Cloud "all this manifested in in a more larger subtle universe of ultimate mass quantity. Due to impulse rate and the effect of growing against a foreign entity dark matter - the cosmic eggs mass grew until it reached a critical density against the dark matter that kept it in position and power. Dark matter being flexible by its own nature - reached a critical point that the fabric began to spin its fabric impressed chemistry into motion and like a car hitting a block wall it scattered bits and pieces everywhere. The believed negative repulsing of dark matter at this time in Quanta Physics is believed to have reached 10 -480g/cm^2. (10 -960g/cm^2)x *pi*

Light Speed Technology

American physicist Albert Abraham Michelson, was the earlier physicist that measured the speed of light. Michelson devoted his entire life to devising ingenious methods for clocking nature's fastest phenomenon, the entity of light itself. Despite high sensitivity experiments such as the Michelson/Morley Experiment that measured light in vacuum. The results of his experiments always finished with a photo-finish. Especially, the experiment of measuring light between two light beams. Michelson and Morley's contributions to science were well achieved with the works of Albert Einstein in that era. But other phenomenon activities about this planet were also taken seriously. Some scientist contended that it is the Coriolis force that makes drain water in a bathtub spin clockwise above the equator and counterclockwise below it. It was believed at that time that it retained the activity of multiple invisible forces that accounted for the present of these types of special occurrences.

Observing this planet as a single frame of reference we can say that things observed like the rotation of the earth happen in single frames that it is certainly the earth that is rotating and not the starry sky. It was the imagination of Albert Einstein that perceived to the idea that there existed no real meaning to this question. "Who is really moving" the earth, the stars at night or the universe itself altogether. This was the building block to his new theory.

Relativistic mass increase instituted the first barrier to FTL travel. If you tried to accelerate a body up to the speed of light, it gained mass as it approached the Einstein limit. At the speed of light itself, the body would have infinite mass - immovable no matter how large the applied force. Near the speed of light, most of the energy put into a body goes into increasing its mass very little into increasing its velocity. Because of the relativistic mass increase it was impossible to accelerate a particle to a velocity equal to the speed of light at least noted in that time and era. Einstein's limit procured the notion in the earlier twentieth century that subliminal speeds could not be achieved by brute force due to this analogy in his theory.

About this same time of Einstein's discoveries on relativistic mass limitation the Stanford linear accelerator was designed to speed (electrons) to near light speeds. To get the SLAC electrons moving at their top velocities (99.99999992 percent light speed. The accelerator had to be 2 miles long and cost near $300,000,000. Dollars.

Another limit postdate theory that prevented faster than light space travel in theory was called COP the Causal ordering postulate, it was not a part of relativity but was an extra assumption concerning the presumed nature of causal relations in space-time. The extra assumption of COP provided unlike the relativistic mass theory, a virtually leak-proof barrier to FTL travel and/or FTL signing. To understand COP was aligned by measuring the contraction of matter as it approached near light velocity and accounted with the Lorentz transformations act on the space-time analogy. As we read further I must tell you that " The Quanta Physics Theory " that has now been established in its publication of theory that mass increase and FTL COP orders have been re-investigated and a detailed alternate theory has been developed by the research of this author Rodney Kawecki 2007.

"Time is simultaneous because "e" in this universe is structured in a medium "e" orbital universe slow-drag effect that sustains and procures data in an orbital circulatory spin of the universe's axis. Energy "e" is therefore strung along at a light velocity dwelling that maintains a physical storage of the info or data made without the universe region as a whole and independent to its entities.

QUANTUM LIGHT SPEED HYPOTHESIS

Reviewing what mentioned in Rodney Kawecki's book entitled "THE SUPERTELLIC ELECTROMAGNETIC-GRAVITATIONAL UNIVERSE TECHNOLOGY THEORY" what is called " The Quantum Light Speed Theory' I will explain in more detail. The best evidence to superluminal space travel theory maintained in today's modern physics that shows physical evidence to this hypothesis I am led to what is called "DIRAC'S SEA".

Research on the possibility of faster than light space travel can well be addressed with the work of Paul Dirac physicists at Cambridge University in 1925 Dirac's work on anti-particle behavior at this time show that what is called THE DIRAC SEA equations act relative to the effects of an external force as Albert Einstein's work on relativity and light speed doctrine this external force of gravitation acting on all particles and matter on this planet.

Dirac's work maintained that when the positive energy electron particle is in a ZPE state or field vacuum it resides at rest. This evidence refers back into the Quanta Physics Study and chapter of Quantum Light Speed Hypothesis in the manner that infers to the Interstellar space field realm. Dirac's hypothesis on particle behavior and referring to the activity of electrons and the electron antiparticle raises the issue to faster than light speed possibilities that in Quantum Light Speed Theory infer the possibilities to time travel within the standard model.

The PE electron (positive electron) is in its lowest energy state when it is not moving. At this point a PE electron has only rest mass energy equal to mc^2. The faster the PE electron moves the more energy it acquires over and above its rest mass. As a PE electron's momentum increases it moves up the energy ladder.

When a NE electron (negative electron) is not moving, it resides in its highest energy state with rest mass equal to $-mc2$. Presumably, a particle with negative mass would fall up in a gravity field, but no such particle has ever been observed. As an NE electron moves faster, it actually loses energy and moves down on the energy ladder. {Q.L.S.T. (If you ad energy to a neg. electron it will slow down)}.

The negative electron can lose energy without limit continually increasing its velocity. {Q.L.S.T.(While its opposing particle the PE resides continually at rest during this time in phase)}. Because of electromagnetic positive and negative charge interaction, a electron by emitting a photon of light carries away the extent energy between any of the two acting state. This hypothesis in Quanta Physics asserts that while the particle mode is in negative state velocity that its counter-part PE electron resides at rest assert the emitting of positive energy flow. Referring to speed of light in vacuum one can assert that matter or massive particle components "e" resume an at rest characteristic of the phase from phase one velocity. Also that superluminal space travel in vacuum space the NE electron asserts dominion stability when engaged against Quantum Light Speed Force...or engaged torque speed entailed in a zero point energy vacuum space field condition space. With these qualifications when we review what relativity asserts as infinite mass we can resort to this resolution of particle phase of the Quantum Light Speed Hypothesis explained here.

We must include the facts on infinite mass theory with negative vacuum space field theory to cancel out the infinite equation. We discover that the Quanta Physics Study illustrates n empty space circumstance field grade thereby not enabling the idea of a negative energy space type field space. The Dirac Sea scheme that the vacuum is empty of positive-energy electrons assumes to be correct allowing the resolution to a cosmic solar wind activity and sound effect in space but he also claims that the Interstellar space field is filled with an ocean of negative electrons.

This assertion in Quanta Physics refers back to the idea of what dark matter is or contains for its components. The idea that *Kaweckian Space* is a lot more less than even that as an equilibrium which negative density is the component in itself for a space field chemistry definition also shows that it is not a material quantity of the space field that is infinite but the dark matter it arose from and or by it. This idea also opens the realm to multiple disk type universes throughout infinity.

The Dirac examination on antiparticle and or particle behavior illustrate particle that will travel through empty space at innumeracy speeds that resume and are superluminal in characterization but can be accounted for when we include the negative counter velocity part space field density which in Quanta Physics is increased to 10 -460 for the beginning of the universe's birth that is accounted by the torque velocity exclusion of gravitation in the E=Mc2 equations the negative density of the dark matter space field at the time of the big bang event. It also claims that the space perceived not empty is filled with allusive passing particles negative and or positive in nature. That space is real not all empty. Quanta Physics feels that empty space is just a highway for allusive deities like particles but isn't the playground for special time traveling activity as physicists express it to be. The nature of Dirac's Sea is the existence of an unknown type element that fills the nature of empty space. It is un-similar in nature to singular deities we see in the positive scenery.

Alcubierre of New Mexico University, believed because the dark matter terrain may be positive energized by nature as Albert Einstein asserted in his theories that creating this space-wave meant forming the Casmir Plates, that form a negative energy resistances bubble inside close to the center of the wave. The universe would have to be filled with these naturalized particles to create the warp bubble. For such a mechanism to exist it would have to be man-made and not perceived through the nature of how the universe an space acts as an empty element vapor type. Space is subtle and pushes towards the outer rim of the vapor cloud. Celestial curves are formed from the orbiting planets and celestial galaxies and nebulas where the universes orbitration pushes hard enough at light speed that peaks and bottom curves are formed by mobile celestial spheres. A place where not even light can penetrate the twisting fabric nor could a spacecraft maneuvering the realm.

Today's principle closest nature to achieving light speed in a laboratory

Another marvel-like example of FTL motion is the electron beam in an oscilloscope or television picture tube (this accordance with the literature of Nick Herbert PhD author of Quantum Reality). Like the searchlight beam or flashlight beam the writing speed of an oscilloscope speed of which the luminous dot moves across the screen in principle, can exceed the speed of light. In practice the fastest commercial scope, the Tektronix 7104, achieves a writing speed only 60 percent light speed. By simply doubling the length of the 7104's display tube, Tektronix engineers could build a superluminal oscilloscope whose spot velocity would be 120 percent light speed. Some experiments of this scope probably already exceed the light barrier and are kept in commercial and independent privacy acts and documents to be hidden from public domain.

Facts about the Disk Universe

When all the positive energy aspects that reserve positive charge or reflect ' quanta burst ' - it product is a ' non-energetic black fabric. Virtual particles that 'jump' in and out of dark matter may be the missing link to how the universe evolved from a primitive state or embryo. Virtual particles are deeply embedded in dark matter quantity.

John Steward Bell's idea published in a short-lived journal called ' Physics" indicates that a universe haven begun with a force greater than light speed - says that superluminal connections in the universe and its nature should exist. It wasn't until Hubble discovered that galaxies were traveling apart from each other that he along with Albert Einstein to be more specific that the universe was expanding at the speed of light. This idea was later reviewed by Hubble who measured this separation between the stars and galaxies and developed the Hubble Constant.

Quanta Physics has reviewed this evidence and concludes that the universe now orbits as a circular single disk - retaining a oscillating and close to equal orbit orbitration direction. The universe thereby is orbiting in a single forwards direction at the speed of light so Einstein was right about it moving at the speed of light but not expanding. The expansion is found by Hubble and as the disk orbits its expanding or widening at the Hubble constant. Thus we can say that the universe is constant with two equations.

Called "The Oscillating Universe" its circular disk spin creates a center fabric-hole for resistances in its circular motion, resistances is served at its center creating a central vortex or fabric-hole center. Reviewing aspects about the big bang we discover that dark matter is flexible in nature which conveys the possibility of an inflationary period at rapid greater than light speed eruption. The fabric-hole is actually created due to the confined spin and limiting the critical point of the explosion that occur at time + zero where its effect are formed by angular-momentum. The gradual weight indifference as the material of the big bang cooled over-time causing the universe development to set at an angle positive in spin as the explosion was subsided by inner and embedded dwell. (Inflation)

The Kawecki Universe spins at the speed of light maintained due to the weight strives against the subtle dark matter around it and setting the universe in constant motion with nothing else interfering. At the universe's edges inflation caused the circulatory ends to curl upwards on top and downwards at the bottom of the orbiting disk. The big bang caused a permanent warp in dark matter that the universe today and then recedes from.

CHAPTER SEVEN

GRAVITY BY ATMOSPHERIC ENERGY PRESSURE

To understand gravitation we should look at the earlier decade of theories that exist and how the equivalence principle Einstein developed in 1915 explains it, imagine how it would be if gravity worked like other forces. If gravity were like electricity, for example, then balls with more charge would be attracted to the earth more strongly, and hence fall down more quickly than balls with less charge. (Balls whose charge was of the same sign as the earth's would even "fall" upwards) There would be no way to transform away such effects by moving to the same accelerated frame of reference for all objects. But gravity is "matter-blind" — it affects all objects the same way. From this fact Einstein leapt to the spectacular inference that gravity does not depend on the properties of matter (as electricity, for example, depends on electric charge). Rather the phenomenon of gravity must spring from some property of space-time.

The equivalence principle was properly introduced by Albert Einstein in 1907, when he observed that the acceleration of bodies towards the center of the Earth at a rate of 1g (g = 9.81 m/s² being a standard reference of gravitational acceleration at the Earth's surface) is equivalent to the acceleration of an inertia moving body that would be observed on a rocket in free space being accelerated at a rate of 1g. (1g (g = 9.81 m/s² x 10 = 98.1 m/s²% [1,000G])

186,000 miles per second

18.5 miles per second

Earth Rotation

subjected sum

1,000 gravitons

Einstein stated it thus:

In physics, the center of mass, of a distribution of mass in space is the unique point where the weighted relative position of the distributed mass *sums to zero*. The distribution of mass is balanced around the center of mass and the average of the weighted position coordinates of the distributed mass defines its coordinates. Calculations in mechanics are simplified when formulated with respect to the center of mass.

Einstein quote suggests to us that it is the center of mass of the celestial that commonly keeps the planetary sphere's separated and positioned. That space curvature is just space-time activity due to the positioning bodies weighed. It does not characterize the fact that the universal axis and velocity of gravity measures as the universe spins. In Quanta Physics Mechanics, the celestial's are engraved positioned by the universal axis and its speed. Looking at the dominion star curvature without the axis spin the celestials would fall into the dominion sphere, looking at our own solar system as it orbits the sun.

In the case of a single rigid body, the center of mass is fixed in relation to the body, and if the body has uniform density, it will be located at the centroid. The center of mass may be located outside the physical body, as is sometimes the case for hollow or open-shaped objects, such as a horseshoe. In the case of a distribution of separate bodies, such as the planets of the Solar System, the center of mass may not correspond to the position of any individual member of the system.

Gravitation, or gravity, is a natural phenomenon by which all physical bodies attract each other. It is most commonly experienced as the agent that gives weight to objects with mass and causes them to fall to the ground when dropped.

Gravitation is one of the four fundamental interactions of nature, along with electromagnetism, and the nuclear strong force and weak force. In modern physics, the phenomenon of gravitation is most accurately described by the general theory of relativity by Einstein, in which the phenomenon itself is a consequence of the curvature of space-time governing the motion of inertial objects.

From a cosmological perspective, gravitation causes dispersed matter to coalesce, and coalesced matter to remain intact, thus accounting for the existence of planets, stars, galaxies and most of the macroscopic objects in the universe. It is responsible for keeping the Earth and the other planets in their orbits around the Sun; for keeping the Moon in its orbit around the Earth; for the formation of tides; for natural convection, by which fluid flow occurs under the influence of a density gradient and gravity; for heating the interiors of forming stars and planets to very high temperatures; and for various other phenomena observed on Earth and throughout the universe.

Scientific revolution Modern work on gravitational theory began with the work of Galileo Galilei in the late 16th and early 17th centuries. In his famous (though possibly apocryphal experiment dropping balls from the Tower of Pisa, and later with careful measurements of balls rolling down inclines, Galileo showed that gravitation accelerates all objects at the same rate. This was a major departure from Aristotle's belief that heavier objects accelerate faster. Galileo postulated air resistance as the reason that lighter objects may fall slower in an atmosphere. Galileo's work set the stage for the formulation of Newton's theory of gravity.

The effects of gravitation are ascribed to space-time curvature instead of a force. The starting point for general relativity is the equivalence principle, which equates free fall with inertial motion, and describes freefalling inertial objects as being accelerated relative to non-inertial observers on the ground. In Newtonian physics, however, no such acceleration can occur unless at least one of the objects is being operated on by a force.

In the decades after the discovery of general relativity it was realized that general relativity is incompatible with quantum mechanics. It is possible to describe gravity in the framework of quantum field theory like the other fundamental forces, such that the attractive force of gravity arises due to exchange of virtual gravitons, in the same way as the electromagnetic force arises from exchange of virtual photons. This reproduces general relativity in the classical limit.

The strength of the gravitational field is numerically equal to the acceleration of objects under its influence.

In December 2012, a research team in China announced that it had produced measurements of the phase lag of Earth tides during full and new moons which seem to prove that the speed of gravity is equal to the speed of light. The team's findings were released in the Chinese Science Bulletin in February 2013.

According to albert Einstein and relativity gravity is caused by massive objects (planets, stars etc.) displacing space itself, like (to use Einstein's metaphor) putting a bowling ball on a mattress. It displaces and distorts the mattress so if you were to roll a ball alongside the bowling ball it would be pulled toward it.

Gravity according to Newton is the potential energy between two bodies proportional to the masses of the bodies and inverse to the separation to the bodies.

$$E = -GmM/R$$

Einstein's view that gravity displacing space is mistaken the deflection is shown by Newton, tan (g) = y/R= 1/2g t2/R = 1/2gR/c2 =8.1667E-6 , where g is the Earth's gravity g=9.88 and R is the distance to the sun 150G m.

Einstein's "Equivalence Principle" asserts that a gravitational field cannot be distinguished from a suitably chosen accelerated reference frame - essentially because we cannot distinguish between the reciprocal cases of space-time accelerating THROUGH US (gravity), or our own acceleration through space-time (as in a rocket ship). Hence the equivalence between inertial and gravitational mass - classically recognized by Newton as the equivalence between inertial resistance and gravitational "weight", but not understood. Einstein said that gravity was acceleration and space-time curvature and were both equivalents.

Quantum field theory depends on particle fields embedded in the flat space-time of special relativity. General relativity models gravity as a curvature within space-time that changes as a gravitational mass moves. Historically, the most obvious way of combining the two theories in such a way is by treating gravity as simply another particle field. Theoretical physics harmonizes the theory of general relativity, which describes gravitation and applications to large-scale structures (stars, planets, galaxies), with quantum mechanics, which describes the other three fundamental forces acting on the atomic scale. This problem must be put in the proper context, however. In particular, contrary to the popular claim that quantum mechanics and general relativity are fundamentally incompatible, one can demonstrate that the structure of general relativity essentially follows inevitably from the quantum mechanics of interacting theoretical spin-2 massless particles (called gravitons).

While there is no concrete proof of the existence of gravitons, quantized theories of matter may necessitate their existence

$E=Mc^2$" redirects here. For other uses, see $E=Mc^2$ (disambiguation).

Principle of Relativity

In physics, mass–energy equivalence is the concept that the mass of an object or *system is a measure of its energy* content.

Chapter Eight

The Fabric of the Grid

In space, in outer space it is not all free fall coordinates that measures how fast a ship can travel at. It is divine intervention flight in zero point gravity and how the vacuum is. The deeper the area of the vacuum the faster a spacecraft might travel. There is most likely to exist deeper regions of space regions and planetary sections that free fall backed up by propulsion accelerations will push a ship to a most extreme velocity.

Over the decades the theory about gravity also researched by Johannes Kepler born in December 27, 1571 – November 15, 1630) was a German mathematician, astronomer and astrologer created a model of the solar system which he was held for treason for his idea. When we review that facts about gravitation at the celestial scale amongst the planets and stars we have to look at the whole picture. The universe spins slow dragging dark space with it as it rotates. As a whole it is the weight of the galactic matter that space forms into a rotating disk plate. It creates a virtual fabric that allows the heavenly bodies to become impressed into in deep impressions in the fabric. It's like the universes spin forms a non-penetrable surface relativity calls the fabric of space when in reality it is the universes spin that forms the surface grid. If the universe did not conform to this analogy disk dominion stars and suns would dominate space pulling in nearby spheres into its realm. But as it is the spin by the universe causes matter to push towards its outer rim edge. We look at the universe in allusive orbitration pulling the galaxies where in total motion it is the weight of the galactic matter that is slow dragged rippling the flat rigid space disk plate. Inside the galaxy's solar systems of planet hard ware retain an orbit due to the swift impression the suns retain on nearby planets but are contradicted by the universes spin. The density of a planet's atmosphere stretches into deeper realms in deeper regions inside the more dense galaxies that exist throughout space. But the universes spin is what causes the alignment of the planet structure that lay inside revolving galaxies, nebulas and clusters of stars.

If the universes planetary matter did not drag along its maneuvering the dark fabric the planets and stars making up the systems would fall downwards into the anchorage of great suns throughout the grid. It is the universes spin that the weight of matter traveling at terminal velocities some faster than others that create a balance in the universe all together.

On a subtle bases size mass of celestial spheres great and small heavy and gaseous are separated by 'similar' weight mass the spin of the universe creates, its energy mass only creates a shield for resistance amongst the bodies all matter is caught within the deep impressions of the dominate planetary realm we know as galactic celestials. Early physicists leaned down towards the idea that it was energy and attraction that the dominion of the gravity force manifested itself. But the truth of the matter is - like a moving car at is not friction that slows the carriage to stop but the lack of temporal force acting on it. Unless force acts as a continuum to moving objects like cars, boats etc.., they'll all slow from and cease to move.

Free fall in space and abroad earth surfaces venture gains by the lack of thrust. Terminal velocity acts as a balance between an objects 'weight' and its 'velocity'. As one free falls from the sky it is interrupted by the earth's rotation and tension of its orbit. Each second on the clock the earth moves 18.5 miles per second intervals per second a free falling object is pulled by the pressure elements (air, heat, wind rain) bestowed in the atmosphere becoming heavier and heavier as it reaches the surface. As it falls it's caught within the planets rotational track that layers the atmospheres invisible element forces twisting by the planets velocity. Bound by the planets elliptical path none of this change to any much of a degree over time, the second units of a time clock interval seems to be a constant amongst everything that exists throughout space even the pulsar energy star impulse waves.

Space seems to be divided by elliptical plates that each section of space is divided by. The solar system divides its space area from a proceeding existing star system or system of rotation formed by some nearby system of planets outside our system dominion of planets. Elliptical highways are formed between these plate activities throughout the grid. There exist many systems in a galaxy each rotating plate traveling faster or slower than the other. The elliptical path between them act as highway passages for spacecraft flight this discovery of elliptical plates was finally discovered in 2013 by a satellite passing outside our solar system.

There exist two layers of space that exist in empty space field. The zero point disk alignment the planets, stars and galaxy's lay and impress themselves in bending the fabric's upper disk plate or elliptical disk plate as I explained about in the earlier paragraphs. Then there's the empty space field grade lying above the grid. It are the planets and stars that layer the space grid with impressed wells that lay deep within the top of the disk plate. Solar winds form at the dividing caps of curved space and the planet disk plate each celestial implants with its positional force of its weight and rotation. Above the impressed planetary disk plates, like the one our sun has impressed on the fabric galaxies have impressed the greater impressed grid which 'open space' lays above them if the fabric of space acts with the impression of galactic matter than it is not empty because the planets and stars act in its midst. But above this darken element fabric lays an open terrain of empty space that acts as free space and is not occupied by celestial matters.

Free fall velocities and terminal speeds measure at all points and areas of un-occupied space. The free fall of universe galactic planetary matter as it orbits allowing momentum between the grid and empty zero point gravity (grid gravity) from which our solar systems rotate and orbit in and are formed caught inside deeper dense perimeters like in our Milky Way Galaxy.

History about the gravity force researched over the centuries have led scientist like Isaac Newton and Albert Einstein amongst only to mention a few to believe it is 'energy' or the 'attraction' of that energy inside matter that creates the gravity force. But is the impressed force of the universes action it takes amongst its celestial sphere and galaxies that its layered zero point empty space element raises out from.

Neil Armstrong in 1963 headed an experiment on the moon by dropping a feather and a hammer at the same time as he sat sitting at his spacecraft's upper step form. He performed this experiment to challenge Einstein's theory about object mass and gravity. They both fell to the surface at the same speed. The hammer no faster than the feather but the fact remains that on the moon that retains as little as eleven per cent g-force that the lack of planet element existed on the moon's surface terrain. That eleven per cent of 1 g^2 comparison to earth's gravity equals about 1 per cent close to nothing. There existed no chemical elements as earth, air, wind or water on the moon. Its energy star content is near zero gravity as Armstrong could jump fifteen feet through the moons atmosphere only falling due to his measurable weight. What the experiment did prove was that the space that layered the moon's surface that retained no atmosphere at all – showed that space is a measurable zero point gravity realm even if it surrounds a desolate satellite like the moon.

CHAPTER NINE

WHAT FORMED EARTH AND A LITTLE ABOUT THE LIGHT

The speed of light constant in space is less than the speed of light in a vacuum due to dust and gas in space. The speed of light is slowing down in space, and this gives a far different time of creation. Light travels faster in space than in air, gas, water, glass, diamonds or rubies. The undeniable fact about the speed of light is that it is not constant. Scientists have measured the speed of light in space at 186,282 miles per second (a partial vacuum, we will discover) and in the earth's atmosphere a little less, in water at 140,000 miles per second, the speed of light in glass at about 125,000 miles per second, in diamonds at 77,600 miles per second, and in rubies at only 190 feet per second. They know that the speed of light is slower in air than it is in a partial vacuum; that it is slower in water than in air, and slower in glass than in water. All their other conclusions about red shift, Big Bang and the speed of light are based on space being a perfect vacuum and light having a constant speed of 186,282 miles per second in space, or 299,792,458 meters per second.

Ten years later in 1992, I realized that the Hubble space telescope and others show space filling with gas and dust. Many enormous clouds of dust and gas in space are so dense that they completely obscure the light from any stars behind them. Hubble's Bubble is a good example. This enormous cloud of gas six light-years across is in the constellation Cassiopeia. Dust and gas decrease the speed of light.

Does the decrease in the speed of light conflict with the statement frequently attributed to Albert Einstein that the speed of light is constant? Not really. Einstein's theory of special relativity assumes that the speed of light is independent of the velocity of the light source.

Very bright, exploding stars are called "supernovas." If starlight, apparently from a supernova, were created en route to Earth and did not originate at the surface of the star, then what exploded? If the image of an explosion was only created on that beam of light, then the star never existed, and the explosion never happened. Only a relatively short beam would have been created near Earth. One finds this hard to accept.

Instead of explaining this abundance by supernova, Tang and Dauphas propose that a massive star (perhaps more than 20 times the mass of the sun) sheds its gaseous outer layers through winds, spreading aluminum 26 and contaminating the material that would eventually form the Solar System, while iron 60 remained locked inside the massive star's interior. If the Solar System formed from this material, this alternate scenario would account for the abundances of both isotopes.

"In the future, this study must be considered when people build their story about solar system origin and formation,"

The most widely accepted model for the origin of the Solar System is called the nebula theory. Most generally phrased, the theory states that the Solar System condensed from a large, lumpy cloud of cold gas and dust. This idea was first in the late 18th century by two Europeans, Immanuel Kant and Pierre Laplace. Extensive observations since then have confirmed that the nebula theory is the best explanation for the origin of the Solar System. All theories are subject to refinement as new data is gathered.

According to the nebula hypothesis, the Solar System began as a nebula, an area in the Milky Way Galaxy that was a swirling concentration of cold gas and dust. Due to some perturbation, possibly from a nearby supernova, this cloud of gas and dust began to condense, or pull together under the force of its own gravity. Condensation was slow at first, but increased in speed as more material was drawn toward the center of the nebula. This made gravity stronger, making condensation faster.

The Sun formed from material that condensed in the center of the spinning disk.

The nebula also began to spin counterclockwise, as it conserved the angular momentum of the material drawn toward the center. This spinning made the material around the center of the condensing nebula flatten out into a disk-like shape. Nebulas, at this stage, have at its center a roughly, spherical core, surrounded by a disk. This has been observed by the Hubble Space Telescope. The remainder of the nebula theory is based more on modeling and indirect evidence.

The center of the nebula continued to contract due to gravity. Eventually, pressure and temperatures in this mass became high enough that nuclear fusion started. The central mass became a star, the Sun.

While this was happening, condensation was also occurring in the disk. Gas and dust came together to make tiny particles, which gradually joined with other particles, making larger and larger objects. These objects grew to be several hundred kilometers in diameter; they became proto-planets. The proto-planets had much stronger gravity than the very small particles of gas and dust around them. They began to behave almost like vacuum cleaners, attracting the small particles around them. Proto-planets also collided from time to time. These collisions, plus the "vacuuming" of small particles, formed the planets of the Solar System.

CHAPTER TEN

GRAVITY WAVES, DAYS, MONTHS AND YEARS

What is a *Ligo's* Gravitational Wave?

The Answer:

A gravitational wave is an invisible (yet incredibly fast) ripple in space. Gravitational waves travel at the speed of light (186,000 miles per second). These waves squeeze and stretch anything in their path as they pass by.

The problem with this analysis is who made the assumption gravity waves propagate at the same speed as the speed of light.

A gravitational wave is an invisible (yet incredibly fast) ripple in space.

We've known about gravitational waves for a long time. More than 100 years ago, a great scientist named Albert Einstein came up with many ideas about gravity and space. Einstein predicted that something special happens when two bodies—such as planets or stars orbit each other. He believed that this kind of movement could cause ripples in space. These ripples would spread out like the ripples in a pond when a stone is tossed in. Scientists call these ripples of space gravitational waves.

Gravitational waves are invisible. However, they are incredibly fast. They travel at the speed of light (186,000 miles per second). Gravitational waves squeeze and stretch anything in their path as they pass by.

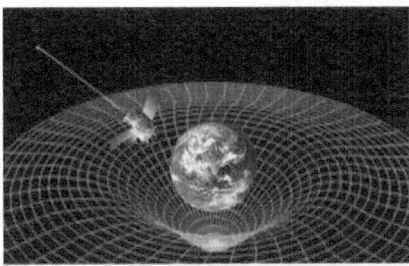

Made with this scientific claim we have discovered that light speed is not the fastest deity known to science. We do have excited repulsive forces in space that unlike those in a planets environmental atmosphere are different than those. In a planets environment gravity is the force of attraction. It slows the acceleration of moving objects and without a constant force to keep these surface objects in motion they will finally stop.

In 2015, scientists detected gravitational waves for the very first time. They used a very sensitive instrument called LIGO (Laser Interferometer Gravitational-Wave Observatory). These first gravitational waves happened when two black holes crashed into one another. The collision happened 1.3 billion years ago. But, the ripples didn't make it to Earth until 2015!

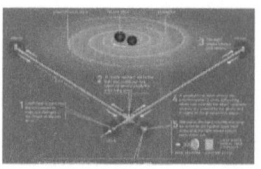 

Ligos   **Gravity Path**

In space, the universe acts as a whole. Gravity waves are made up of gravitational paths from object presents and motion. As the universe expands and gets larger waves are formed from the planetary masses aboard the space field. They are not sectional distortions as relativity stated a hundred years ago. Gravity are waves forming the paths of these objects in space that form them vanish just as fast as they are made If a planet or star were to explode or vanish so would its existence and orbital path. We have the repulsive force. It acts opposite of attraction which most gravity alignments follow. They push away they don't attract and their impulse force acts at about 450,000 miles per second.

Yes, gravity waves ripple space but at a much greater velocity than the speed of light. Matter energy propagates at light speed but space does not. As well the universes expansion rate changes every second of time on the clock. Reviewing the dimensions of this runaway universe will help you understand the exact nature of its process. Energy reluctantly is shadowed by its own continuity. I is not planetary matter that propagates electromagnetic-waves as energy but it is dark gravity that is repulsified by the presence of planetary matter as a non-existent form. A form different of itself and it pushes it away. Path-waves are supported by impulse generated reverse-vacuum activity that happens when it (space) interacts with the presence of planetary matter.

Using the Element Chart, we understand that Dark Gravity has 96 % percent more element than matter-energy. It's the weight of repulsive power that out-weighs energy-mass not the reverse.

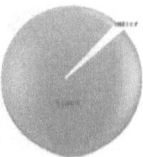

Trying to turn back Dirac's Sea for time travel

A gravity force that acts in the same way as planet gravity but is repulsive. The space gravity is a force that accelerates moving objects in space to a zero mass-weight performance. An object able to accelerate to 500 miles per hour is actually traveling 1,000 miles per hour and so on up the chain of propulsion capacity.

Unlike planet gravity the force acting on a moving object is slowed down by fifty percent its initial accelerated force. This is due to attraction of the planets gravitation. Where by in space a moving is not slowed down by fifty percent acceleration but is accelerated due to space expansion. So any moving objects acceleration force is enhanced fifty percent.(not slowed by attraction)

When we review aspects to Einstein's view in these cases we discover that his theory to gravity in space stays the same. What attraction does to moving objects in a planet environment happen s also in space? Whereas Einstein's theory on gravitation in space fails based on the circumstance.

We review light velocity as a constant in relativity. As it is believe to be the fastest known deity to exist. But when we review Erwin Hubble's theories we discover space is enhanced and is accelerating even faster than light. What this does to the theory of gravitation is changes how we view gravity in space. As a repulsive force a ship able to travel close to the speed of light (e) its acceleration is enhanced by the space gravity force and the ship travels 372,000 miles per second. The masses initial weight is accelerated because its not effected by any attraction force as it would be if the ship were on earth. The ship therefore travels its liminal light speed velocity plus another 186,000 miles per second due to the lack of attraction. Push force only means that planetary bodies are advanced in a zero point gravity in space where attraction does not exist. The ships acceleration that would otherwise be attracted causing the ship to slow down doesn't sow down at all but is initially enhanced not only by the lack of attraction forces but is also accelerated by the universes expansion that a push force abides by.

A.E. equation for or attraction with amass capable of traveling light speed is slowed down due to the force of attraction and can only travel 93,000 miles per second under earth's gravity force conditions. In space this same object has the ability to travel is true velocity which is equal to the speed of light 186,000 miles per second. Of course in space this same object now travels its true velocity 186,000 miles per second. But due to the universes space expansion push force it travels faster.

These facts are shown to equals 2.7c. 450,000 miles per second. This is equal to the velocity of light plus 1.7c more due to space universe expansion.

Albert Einstein, official 1921 Nobel Prize in Physics.

The most powerful gravitational waves are created when objects move at very high speeds.

But what I have just explained to you show that in space gravity waves are propagated through events that a new scale of gravitation is measured by. Instead of a maximum velocity or universal constant that force is equal to the speed of light it now follows the forces created by merger events that might show the existence of gravity waves exist in a greater affection of space interference than thought.

Do gravity waves act repulsive and propagate stretching outwardly or do they propagate inwardly against the expansion velocity

When the universe expands it does by pushing outwardly. Like a balloon equally at all sections and all at the same time. Its put in spin by the repulsive of normal idle gravity particles that wish to be at rest, accelerating a push on the matter at an impulse rate that causes all matter to axis and spin and all the same way. Ripples in space are conditioned by these forces. If these impulse wave-particles are accelerated by the motion of its repulsive impulse to matter they may propagate at light velocity but this is only the beginning. And it is usually that shock waves and forces that induce g-waves or ripples in space diminish over a specific length of planck-time.

If it is true that the fabric of space surface according to relativity is truly the first point of gravity in space and the density relative to the velocity is the speed of light. And the *space bubble* which is expanding the universe is the source of that gravity and it measures faster than the light speed than that velocity should be the space continuum speed limit. If the blue space bubble structured velocity is 450,000 m/s, even if we were to consider m° as z or zero, (E=½mc) the velocity under Albert Einstein's conditions than measure 225,000 m/s. If the gravity in space is m°, z or zero, believing the space above the grid (string membrane) is zero (ZPE) (ZPG) the speed allowed measures 450,000 m/s as calculated by this author in this book is the *Physics*

The Quanta Physics Theory

The figures are shown here:

93 billion light years (*flat* universe) *astronomy dictionary* Albert Einstein GRT

Round-about three-dimensional mass – *a multiple of pi.*

279,000,000,000 light years (^{2}c) *pi*
       3,000,000 light-years (minus a zero point for gravity) to maintain a substantial rate of velocity)

10,000 x 45 miles a sec (46.2 plus or minus 1.3 miles) = 450,000 miles a sec

Without using light frame measurement intervals"

(450,000 m/s) divided by L.S. = 372,000 m/s plus 78,000 m/s equals Speed without Limit
Equals: 44.7 miles per second every 3 million light-years
{450,000 m/s @ 99.999999 = 372,000 m/s} $E=mg^2$
(2.7□)

Distance to Neatest Star

14 light years to nearest star = 8,189,475,840,000 miles away
One light year = 31,449,600 seconds = 584,962,560,000 miles
$E=mg^2$ @ 372,000 m/s

Today's modern calculation to the nearest star using relativity theory
Actual 7 year journey to nearest star

Days, months and years to neatest star
8,189,475,840,000 miles to nearest star
18,198,835 seconds @ g^2
303,314 minutes @ g^2
5055.2 hours @ g^2
210.6 days (to neatest star)

....

Using The Quanta Physics Theory in retrospect [a difference in 'G' matters.] {2 x L.S. plus .7c}
= 7 months (to neatest star) @ g^2

The universe has an expanding gravity force with a velocity equal to 372,000 m/s plus 78,000 m/s or to be more exact in numbers: 450,000 miles per second is the accelerating speed of gravity a carrier force that continues to grow greater over time. All this in accord with the common standard theory about the universe as we know it there are two different origins about the position of the universe. The first is its matter oscillates inside an expanding bubble formation. The second is that matter oscillates outside the inflation of growing pressure. We will be deciphering the universe oscillating outside the bubble whereas the standard theories about the universe all maintain their sufficient is operating inside the bubble formation. This indifference between the two universe theories of sufficiency and universe structures science have only researched using the inside bubble theory over the decades whereas Quanta Physics is the first theoretical science that describes the universe outside the inflating universe bubble pressure and expansion observing all the common laws of physics which render The Quanta Physics' Theory a new theory of advance physics.

We look at two parts about the universe little has been written about. The first is the earlier fate of its origin and the second where the speed of light stands between both of them. We look at the earliest avenue about the universe's origin and discover that in part of its collapse as a matured cosum the universe underwent a closed space barrier explosion. As part of its collapse the universe expanded in for the most part of itself to a degree it expanded beyond the density the universe is measured by today. As a closed eruption there existed no escape of the pressure that rose from the collapse. Because of the high degree of infinite pressure evolved and with breakage in the pressure built by the collapse and the pressure raised by it the strength of the expansion caused by the cosum collapse not only put everything in spin but also allowed only one way for the pressure to escape and that was to make concentrated.

The young universe raised by density due to the raise in the pressure had no direction but to shrink backwards in on itself as it did when it collapsed. Nothing escaped the realm of the cosum bubble and its early mastoid.

Reviewing these two existing parts playing in the early universe's origin we for the second question have to answer the question about the speed of light. The speed of light seems to be an optical about the universe when it was first born to even today 13.8 billion years later in its history. The speed of light optical seems like it can't be reckoned with. Albert Einstein famous quote nothing can travel faster than light seems to useless to try and change its relativity. But the fact remains the speed of light at the earliest time of the universe's young origin during the expansion haven stretched and cause space to warp due to the flexibility maintained in relativity that classifies its character condensing of matter during this young time and birth of our universe space was not originally tough enough to withstand staying warped by the universe's creation. Space stretched and curved back on itself as relativity predicted in 1915 in general relativity theory says it did. To this effect space warped and shrunk back on itself allowing the physical nature of its flexibility as we know in today's modern physics. And by developing this way the speed of light haven condensed matter into the smallest of its extreme molecular structure formed. Everything we know as matter cannot travel faster than light. It is only that the warp age that space underwent doing creation that expanded and shrinkage that concentrated matter that space co-moving was warp enough and beyond its original fabrication allows for an open source speed limit.

What physicists do know about the wit of the universe and its creation that warp age of the vacuum fabric allows for a faster speed limit than that of co-moving planets, stars and even galaxies undergo even when they are set in motion by the wheel of inertia. From a subtle point and view space is motionless co-moving as a whole piece of the pie and everything in it is in motion due to an axis measurable by a rounded universal edge an edge whereas in a flat universe type doesn't exist. Event horizons as we observe them on the different planets like are planet earth are seen by the overview of the planets spin and is continuous. The universe shaped due to its existence inside a closed avenue of empty space was shaped with no means from the release of the built pressure created by its collapse nothing escapes nor was there any way for chemistry of its origin to escape out of the space barrier that surrounded it.

Everything in the origin of the universe was built by a greater force than what we perceive in the speed of light as a universal constant. The warp age that occurred during its creation is what allows the universe and all its matter to expand into as it is by the warp age that space is expanding from. Matter that reformed into a material mass that cannot escape the Einstein limit has nothing to do with the space fabric warped in the venture.

Flexible to the degree a ship having the capacity to do so can travel faster than the celestial substance that doesn't attain a difference in its co-moving velocity.

The universe created from a single point as the standard theory explains it has been expanding and growing bigger since first sight of the anomaly. It is in this sense what allows that the galaxies are expanding away from each other and at a greater rate of speed than the velocity of light. As the universe expands the space expands along with it and according to the facts the space gravity is expanding faster in the forefront expanding edge we know as space.

It is space that even to this day and age of the universe that continues to fill in the cracks and curves it created in the beginning. By the weight the celestial planets, stars and galaxies tug boat through space by it is the space at the horizon end that is in expansion space that is moving faster than the matter that was formed in its way. From the single beginning origin in space the expanding universe widens the space between its galaxies that also widens the empty space between them. The further the object of the energy between two masses the least it is that any energy exist at all and is how travel at velocities faster than light can be weighed. (zg)

Hubble coordinates of the space foundation are the same that were created by Albert Einstein in the early twenty-first century. The factors for elevation in the modern physics science are led only by the acceptation in the rule. In a space field hence that is characterized as a repulsive carrier force there exist no acceptation that it is the force of space gravity that weakens the walls around the bubble that are universe is shaped by. It is the same force that created the cause and effect in the cosum's collapse. In a closed universe where all the matter existing in it is too small of an amount for celestial gravitation causes to be formed for a universe it might expand forever.

According to Albert Einstein the speed of light was rectified when the universe was smaller than the point at the tip of your finger. As the universe grew it became more and wider between the galaxies. The speed of light changed also everything we thought we knew about the universe was figured wrong. The gravity mass as we knew it grew a part over time and the energy it possessed weakened. The speed of light was no longer measured in the manner it was long before centuries earlier because the matter the mass energy generated from only existed in the dead of space.

The Age of Universe is measured by the Age of the Oldest Planet

The Big Bang Theory is the leading explanation about how the universe began. At its simplest, it says the universe as we know it started with a small singularity, and then inflated over the next 13.8 billion years to the cosmos that we know today.

Because current instruments don't allow astronomers to peer back at the universe's birth, much of what we understand about the Big Bang Theory comes from mathematical formulas and models.

Methuselah Star

Image shows the oldest star with a well-determined age in our galaxy. Image:

© Digitized Sky Survey (DSS), STScI/AURA, Palomar/Caltech, and UKSTU/AAO

Called the Methuselah Star, HD 140283 is 190.1 light-years away. Astronomers refined the star's age to about 14.5 billion years (which is older than the universe), plus or minus 800 million years.

In the Quanta Physics study and book you are now reading. It has come to the table that the *Methuselah Star*, is older than the universe.

In The Quanta Physics Theory, Kawecki explains a gap in the time coordinates of the aging of time and space physicists use in measuring age data throughout the cosmos. In theory, he recognizes that just after the big bang event the reformation of matter re-blemished back into mass show the chunks of matter immediately reforming into galaxies throughout the cosmos system like they are today but at the earliest time of creation. These gigantic chunks of reinvented matter spawned into separated deities were young enough that their energies were close to if not almost equal with the earliest mass they were spawn from.

What this means is the energetic content of the earliest galaxies energy volume was enough that in the midst of re-formation over a specific length of time later exploded leaving behind erotic black hole formations at the midst of them all.

When scientist explain the origin of reformation after the big bang there is not recollect of stages that may occurred. In The Popcorn Expanding Universe Rodney Kawecki has published a 250 page reconstruction of what might be loss information that has led scientist to discover older planetary model's in space but with no explanation why they exist with such a greater age volume than is known to them.

But in a universe as big bang as it suggest expanding at a vast velocity as it is and without the counterplea to the suggested energy-mass of a three-dimensional deity estimates might be at loss with no explanation. The length, the mass and the universe's age all come into play when Kawecki speaks about a greater expanding universe as he does in his books.

Though the expansion and loss of stress dark matter has as a repulsive force against the galaxies the bundling of tension dark matter puts on galactic masses prove to show a relief of tension as the universe expands at its continuum. The loss of tension also relieves a planetary bodies raise or fall in energy readings leaving behind a false reading. How long matter has resided in space reluctant to later expansion readings that may vary over time has relieved us from what the universe's age might be over 10 billion years ago and now. This gap in detecting the energy volume from an earlier date reluctant to the Hubble Constant might read differently if it were not the distant between expansions that is observed but the energy volumes of the bodies using the same method.

Aside from the intervention about ''gravity waves'' what about ''gravity paths''? As you have just read the universe is expanding and not at the velocity of light as many physicists' believed at the time. New information shows an expanding universe running at a velocity of 450,000 miles per second. When we review ''''Ligo's Gravity Waves'''' what are we looking at waves propagating at light speed as expected or 450,000 miles per second as *Kawecki and Hubble Expansion* may persist? What is history suggesting?

Chapter Eleven

Time Travel

Microwaves are so tiny, and constant, that if we can learn to measure it, we will be able to make clocks that are accurate to .000000000000001, or one-quadrillionth of a second! The problem about time travel is that the faster you can travel the further distance you arrive at from your targeted planet. This is why I pursue the theory for instantaneous space travel as the fastest a ship might travel any time in the far off future the capacity of propulsion rather than the speed of light. Time measured by a clock cannot represent the motion of a planet like earth that its function is in variable with one the depth of the sun star that it is attracted too and two its elliptical path measuring the planets weight and velocity and third the planet rotation itself. All three of these must be taken into consideration because they are all connected as a singularity system.

To target earth with a ship traveling at light speed and to think you can travel into its motional past isn't logical if the earth is connected to the attraction of its dominion star its sun and the résistance of its orbitration around it. The two numeral equations will not work when there exist three or four sums of action. A planet does not slow down just because a ship outside its equation field is traveling at light speed. There's no connection between the two. The drag the sun retains with the earth is stronger adding the earth's resistance to be pulled into its realm. The universe acts as a resistor factor when its orbitrates at the speed light helping planets stay pulled away from dominion stars like our sun.

How does traveling at light velocity change the course of a disconnected planet like earth it doesn't the action is parallel as the earth in itself rotates and orbits the sun and the ship travels the distance of light speed. In the earlier chapter I mentioned how a ship would have to take on the drag of the planets orbital rotation to cause it to rotate backwards but even so - how does this change the arrow of time our universe acts as a director too? See what I mean?

In 1915, Albert Einstein first proposed his theory of special relativity. Essentially, his theory proposes the universe we live in includes 4 dimensions, the first three being what we know as space, and the fourth being space-time, which is a dimension where time and space are inextricably linked. According to Einstein, two people observing the same event in the same way could perceive the singular event occurring at two different times, depending upon their distance from the event in question. These types of differences arise from the time it takes for light to travel through space. Since light does travel at a finite and ever-constant speed, an observer from a more distant point will perceive an event as occurring later in time; however, the event is "actually" occurring at the same instant in time. Thus, "time" is dependent on space. Gravitational Time

An important aspect of Einstein's theory of relativity to note is that he proposed matter causes space to curve. If we pretend that "space" is a two-dimensional sheet, a planet place on this "sheet" would cause it to curve. This curvature of space results in what we perceive as gravity. Smaller objects are attracted to larger ones because they "roll" through the curved space towards the most massive objects, this idea is opposed when we look at the probability that everything arrived from the same entity or the big bang and there y makes everything similar matter thus repels as positive energy repels like poles. The fact is that larger object celestials will attract smaller celestials but only to a specific point of origin nearby it which causes the greatest degree of curvature. Under normal circumstances, the effect of space curvature is impossible to observe. However, in the presence of the extremes of our universe (such as black holes, where a huge amount of matter is compressed into an extremely small volume), this effect becomes very obvious

A second aspect to the gravitational time postulate is that the faster an object is moving, the slower time progresses for that object in relation to a stationary observer. While in everyday circumstances, this effect goes entirely unnoticed, it has proven to be true. An atomic clock placed on a jet airplane was shown to "tick" more slowly than an atomic clock at rest. However, even with the speeds achieved by a jet aircraft, the time effect was minimal. A more solid example can be seen through an experiment performed on the International Space Station (ISS). After the first 6 months in space, the crew of the ISS aged .007 seconds less than the rest of us on earth (the relatively stationary observers) but when you look at the advancement and lack gravitational attachment as the earth observers the strain on the physical body is much less as well. It is true that atomic clocks preform differently in space than on earth but again the lack of gravity is present. Based time travel theory in relativity observes the planet actually slowing down at great velocities but a planet is part of a infinite universe all connected together into a divine mechanism of planets, stars and galaxies that act in occurrence with the universe as a whole. To think a single planet in such a connecting network of stars will actually slow down because of the high pursuit of a ships velocity is strangely unrealistic. The station moves at approximately 18,000 miles per hour in empty zero point space much faster than the range of normal human speeds. Even with such speeds, however, time is minimal unless you approach speeds close to the speed of light (300,000 km/sec.).

The changes Einstein ushered in with his radical theories of relativity resulted in the now ubiquitous E=mc2 equation, which essentially states that matter and energy are interchangeable (this discovery eventually led to the creation of the first nuclear fission bomb). However, Einstein's equations also predicted the presence of black holes and gravitational waves, and were initially excused as inconsequential aberrations; there is now substantial evidence to support the existence of black holes.

You are in an elevator that is at rest relative to the earth's gravitational field. The gravitational force on your body, called your weight, pushes you down onto the floor of the elevator. However, because you are neither going through the floor nor being thrown into the air it follows that the floor must be pushing up on you with exactly the same force. You experience this reaction force as your weight. In an elevator the most striking thing is that you have become weightless. Because of the equivalence of inertial and gravitational mass all objects fall freely with the same acceleration

In space travel though Time dilation does not occur because the universe is the controlling element and the spacecraft is independently driven. Clocks run more slowly in the presence of gravity because they are superseded by the energy content of the planet as gravity field. In space zero point energy is presented and changes the equation of the source of the gravity.

When we review the aspects about gravity we have two resources. First the planetary activity between the planets energy and elements of matter and secondly space where the realm of energy does not exist at all.

Space gravitation also called Celestial Gravity is the action between the planets in a specific system of stars or solar system. Like energy repel whereas negative and positive poles attract each one another. This is a fact of physics which both Isaac Newton and Albert Einstein did not precede too. Einstein's attraction state of the planets arrived from the idea that dominion stars like the sun create a dominion warpage of the space around that causes nearby planets to it to fall or roll towards this is not a state of space gravitation. In turn the universe spins according to relativity at the speed of light and causes everything in it pull towards its outer rim. This balances the immediate position of the planets system motion occurs when they roll resisting the pull or attraction of the free fall towards the dominion star or sun.

Independent planet gravity occurs due to enclosed system of the planet and its atmosphere. Arising elements fill the atmosphere creating a weight content for movable surface activity on its surface. Since like poles repel – on earth the energy between let's say a human body's movement or strike for motion is weighed down by the physical elements it is made of and the energy within those elements. Matter weighs down the object whereas the push of gravity allows for a specific amount of pushing force. This pushing takes twice the energy-weight of the human body to be set in motion.

The material chemistry of the human matter differs from the energy content of the body which generates the energy the body uses to push with. Earth is a closed system therefore the action within its atmosphere is independent with the mass of the planet and differs from the realm of open space.

Until the time of Einstein, mass and energy were two separate things. In the special theory of relativity Einstein demonstrated that neither mass nor energy were conserved separately, but that they could be traded one for the other and only the total "mass-energy" was conserved. The relationship between the mass and the energy is contained in what is probably the most famous equation in science,

Where m is the mass, c is the speed of light, and E is the energy equivalent of the mass.

Because the speed of light squared is a very large number when expressed in appropriate units, a small amount of mass corresponds to a huge amount of energy. Thus, the conversion of mass to energy could account for the enormous energy output but it is necessary to find a physical mechanism by which that can take place.

Einstein himself originally thought that it might be impossible to find a physical process that could realize the potentiality embedded in his equation and convert mass to energy in usable quantities. In the nuclear age, we now know that he was too pessimistic; there are several physical processes that can accomplish this.

The Theory of Relativity and Distinction

The theory of relativity is *sci-fi*, revealed by Albert Einstein, and is separated into two distinct theories: general relativity published in 1915 and special relativity the actual science fiction story of a life time it was published first out of the two in 1904. These two theories are based on the principle of relativity; this principal states that, whatever referential you occupy, the laws of physics will never change? For example, if you are in a train that passes through a train station, whatever you will do in your referential (the train), an observer on the platform in the station doing the same thing you are doing will obtain the same results in his referential (the platform).

Each of both theories of relativity have different consequences, the theory of special relativity is based on the fact that light travels at the same speed (in a void) for every observer. This can seem obvious but the use of this fact in certain situations (notably high speed experiments) gave Einstein many headaches. Special relativity explain phenomena such as time dilation (explained in further articles), the relativity of simultaneity (explained in the video below), and the famous equation $E = mc^2$. E being energy in joules, m the mass in kilograms and c the speed of light in meters per second Energy is therefore equal to mass time the speed of light squared. This equation shows that mass and energy are proportional (since the speed of light is a constant).

But we can also use this equation to explain one of the basic facts in physics: we know that the mass of an object increases when its kinetic energy does too. Movement therefore increases the weight of objects so the faster an object goes, the heavier it gets. This phenomenon is almost inexistent at speeds far from the speed of light but it becomes more and more noticeable as we get close to the speed of light. An object going at 90% of the speed of light will see its mass double, and an object going at the speed of light (if it initially had mass unlike photons and other particles) should technically have an infinite mass. This is where the energy-mass relation becomes useful, the proportionality shows that to reach this state of infinite mass, you would need infinite energy. The space in the universe itself does not contain any abnormal amount of energy, it is therefore possible for an object with mass to reach the speed of light or exceed it.

Einstein's General relativity, on the other hand, is the use of special relativity on Newton's laws of gravity. (The fact that general relativity and special relativity are applied in theory based on a planetary gravity force and essentially the curving of interstellar planetary systems open space is a vacuum by which energy or energy-mass are not affected by thus the light constant speed limit does not apply). The greatest contribution of this theory to the way of picturing gravity is the idea of space-time continuum. Isaac Newton saw gravity as the simple attraction between two objects which have mass, but Einstein saw it differently: according to him, the mass of any object would affect space and time around it. He explained this effect with the unification of all four dimensions: space-time. Space-time is often represented as a two-dimensional grid in space in which objects having a mass "sink", creating a sort of depression around it on the grid, objects with an inferior mass could then "fall" into the depression or be deviated by it. Of course, this is just a graphic representation of the effects of gravity on space-time, to understand it fully, you must remember that the grid is actually not two-dimensional but four-dimensional, which means that all the special dimensions are "bent" by mass, and most importantly that time changes in the depression. General relativity changed the Newtonian way of seeing gravity, it shows that objects that do not have mass can affect and be affected by the curves in space-time (such as light and that gravity is equal to acceleration mathematically.

Subatomic particles are routinely pushed to nearly the speed of light. The momenta of such particles may be thousands of times more than the Newton expression mv predicts. One way to look at the momentum of a high-speed particle is in terms of the "stiffness "of its trajectory. The more momentum it has, the harder it is to deflect it—the "stiffer" is its trajectory. If it has a lot of momentum, it more greatly resists changing course. This can be seen when a beam of electrons is directed into a magnetic field. Charged particles moving in a magnetic field experience a force that deflects them from their normal paths. For a particle with a small momentum, the path curves sharply. For a particle with a large momentum, the path curves only a little it's trajectory is "stiffer." Even though one particle may be moving only a little faster than another one—say 99.9% of the speed of light instead of 99% of the speed of light—its momentum will be considerably greater and it will follow a straighter path in the *magnetic field*. The Theory of Relativity has been projected through source energy trajectory without evidence to traveling through empty space that possibilities have been ignored until now.

CHAPTER TWELVE

EINSTEIN'S IMAGINATION

This chapter will include the imagination Albert Einstein invented to the world in his this on relativity the imagination of his ideas and formulas that include the equations in modern physics and space travel.

Faster-than-light travel

Scientists and authors have postulated a number of ways by which it might be possible to surpass the speed of light. Even the most serious-minded of these are speculative.

According to Einstein's equation of general relativity, space-time is curved:

$$G_{\mu\nu}=8\pi\,GT_{\mu\nu}\,$$

General relativity may permit the travel of an object faster than light in curved space-time. One could imagine exploiting the curvature to take a "shortcut" from one point to another. This is one form of the warp drive concept.

In physics, the Alcubierre drive is based on an argument that the curvature could take the form of a wave in which a spaceship might be carried in a "bubble". Space would be collapsing at one end of the bubble and expanding at the other end. The motion of the wave would carry a spaceship from one space point to another in less time than light would take through unwarped space. Nevertheless, the spaceship would not be moving faster than light within the bubble. This concept would require the spaceship to incorporate a region of exotic matter, or "negative mass".

Wormholes are conjectural distortions in space-time that theorists postulate could connect two arbitrary points in the universe, across an Einstein–Rosen Bridge. It is not known whether wormholes are possible in practice. Although there are solutions to the Einstein equation of general relativity which allow for wormholes, all of the currently known solutions involve some assumption, for example the existence of negative mass, which may be unphysical. However, Cramer et al. argue that such wormholes might have been created in the early universe, stabilized by cosmic string.

First and foremost part when we review the literature of Albert Einstein he asserts one specific formula in his ideas about space which include 'his imagination". One of his famous quotes includes the statement "Imagine you were traveling in a spacecraft and that spacecraft was a beam of light". When we relook at this quote in his theories of both general relativity and special relativity they bring up the conclusion that his whole ideas and baselines for relativity could be based on what "Imagination". Sure, we have yet achieved the mechanics to build a spacecraft that can over exceed the light barrier if a light barrier exists at all. But foremost, I wish to explain the concepts of Einstein's ideas about space and space travel within the confines of 'his imagination".

What am I talking about when I speak about his theories using his imagination?

In the paraphrase I mentioned about space travel that also include all his concepts about space the most important facts rely on the fact for everything he speaks about that in reality is a 'material' fact it is not a 'light' imagination fact as he explains it is. Space and space flight aviation are a real aspect that continues and will continue to taunt our worlds future the question of whether faster than light space travel is possible or rather it is a dead end street are important amateur questions that we should retain answers too. This is the reason for the collection of my books on faster than light species space travel.

The mainstream of Einstein's work on relativity explains a pattern of events and circumstances that "light" stands as the basis for his theories. The fact is we are not trying to send a ship made of some imaginary light beam into space to travel the distance we refer that what we are sending is materially real.

The ideas of Albert Einstein's work is astonishing when you read it – but the facts he implies about mass objects and light beams used to explain his theories are separate entities and not singular. The idea whether a spacecraft can travel to light speed does not depend on the fact of a light more so it relies on the fact of whether or not a spacecraft launched has the capability propulsion wise to achieve that velocity. Because a light or light beam retains a specific speed limit is dependent on that fact that Einstein is actually talking about 'light' itself and not the object or spacecraft that is being launched.

Let's try and take the 'light constant' out of the equation we will use m=pv (mass equals propulsion velocity or value of which I speak continuously about in my collection of books). We replace 'p' for 'e' and we form an equation based on our propulsion ability rather than 'c' components for the speed of light. Sure light velocity exists but it is an individual velocity for a 'quantum particle' independent to itself. We examine the idea about traveling through space on the value of the ship's propulsion velocity that may or may not be faster than light but the fact that space is a vacuum jump speeds are space vessel into the ability to travel much faster than the speed of light under those circumstances. Why because the highway through the cosmos is maneuvered between the mid-field darkness between the stars and the planets throughout space. This field of space retains a zero point energy deity of empty space which planets and stars retain specific distances from and to each other based on their world energy. Furthermore in my books you will understand more about it is not the specific area of warp space that a world obtains its spin but that it is the universal spin that engages the orbital rotations.

Our ship has a clear distance between the stars. Its capacity to travel at the speed of light without the interference of a positively present gravity field does not exist as on earth. Einstein makes his formulas quite clear that his light speed formula is retained within the gravity force activity. The speed he proposes about light is always and will always be 186,000 m/s this will never change why because its 'light' he's talking about not actual space travel of a material spacecraft.

Using the formula and equation $(- g\square)$ is our equation for the vacuum of space.(a pure vacuum of space based on g-forces relative to earth's might recede to a barrier of $(- g^{100})$ Again we are not speaking about space or space curvature that may or may not cause a planetary spin and we are not talking about the depression it may cause. We are talking free open space vacuum atmosphere an area of space that is mostly empty but retains a realm of darkness. We are not talking about the fabric of space as Einstein calls it that our planetary deities lay and position themselves on No – just cold empty space. Is empty space negatively different to the earth's gravitation I would say it is?

Quanta Physics Theory refers to the facts of the twentieth century data banks about space travel and space technologies that include relativity. The facts that illustrate the commonwealth of Quanta Physics Theory though theoretical relies on the fact that future space economics' and free enterprise rely of real facts - facts which include the idea that a spacecraft having the ability to travel at the speed of light will travel much faster in open empty space. Facts that precede the imagination and are based on facts, barriers and boundaries that are common knowledge in today's world.

Another illusion in Einstein's analogy is his misconceptions' about time. His theories about time and time travel assert the quality of space travel with the idea that velocity allows the ability to manipulated time. He explains time as a four dimension. That a ship traveling through space at a specific velocity the speed of light can intermediately change the perception of time of a targeted planet. He uses earth in his examples. Why not Venus or Mars I can't say why but I do know his ideas about space travel and time he explains are intermediately connected and are interchangeable. Is this true?

He explains that a ship is traveling through space at the velocity of light 186,000 miles a second. He also explains that the target earth is in motion relative to the ships velocity. That because the earth rotates at 18.5 miles a second relative to the ship he is driving at 186,000 miles a second that the sequence between the two mathematical figures show that the ship is traveling over a thousand time faster than the earth's movement.

He then explains that the speed limit to acceleration and velocity is the velocity of light. That nothing can travel faster than that because LIGHT which is mass less and because it travels at its physical velocity which is the speed of light that any massive object cannot penetrate light speed. Here's a short illustration about our solar system. Our solar system is intermediately connected with nine planets with the sun as the dominion star they all orbit around. They are all interconnected and in 2013, an earth satellite showed pictures passing by the end of the solar system that conveyed that all the nine planets seem to be interconnected together and create a disk plate that all the nine planets orbit the sun on. So when I say that all the planets are interconnected that's exactly what I'm saying.

So Einstein's ship is traveling through space at light speed. The earth is rotating at 18.5 miles a second a difference of a thousand miles a second. In Einstein's imaginable dream he indicates that the earth's motion slows down to zero when his ship reaches light speed. He targets earth as his formula of indifference between his spacecraft and the earth's rotational speed. His theory on space curvature intervene here when matter depresses itself on the fabric of space he interpretation is that the planet makes space twist and turn in its presence. In this character he implies that each planet, star or galaxy acts intermediately with in itself with space they create they're spin twisters in the fabric. For this reason time can be manipulated by the velocity of a fast spacecraft. The problem is – is what I have mentioned earlier about our solar systems disk plate where all the nine act together Einstein's twister theory fails against this new technology.

Now his ship is traveling at the velocity of light it really retains no real connection with earth. But four dimensions maintain it does. As the ship continues to travel at 186,000 miles second earths physic-time has stopped allowing time to proceed it. In his original theory in special relativity that these actions are not physically seeable that due to his ship's velocity if the earth could be seen it may in reality be invisible but he asserts that earth's rotation would seem to have stopped.

When we review the aspects about our universe we have to look at the whole idea. In doing this we have to look to everything that is involved in the question we are asking about the fourth dimension. When we do – we discover that in all prospectiveness even though the earth does rotate what causes its rotation in the first place. We discover secondary to Einstein's spacecraft trying to slow down out own planet earth that the universe is making a turn of its own. That as a whole all the planets, stars and galaxy's nebulas and clusters involved that they too are spinning. That as a whole it is the universe on its own that is causing this reaction universe wide. So we have to ask the question about the fourth dimension if it at all even exist whose doing the spinning space curvature or the universe as it spins. The answer is the universe – and if you read more of my books on physics you can learn the reasons.

Einstein's ship is traveling 186,000 miles a second the earth in reality is rotating at 18.5 miles a second. The earth rolls around the sun in an elliptical curvature which takes 365 days in all that means it rolls around in each rotation one time 365 times that equals one earth year. We review the prospect about how many planets act together as a whole on our solar systems disk plate and discover again the number equals nine. That's a lot of drag a spacecraft would have to have to slow drag a planet backwards to travel backwards in time. Einstein's ship would actually have to create a secondary rotational spin vortex to change the planets time frame and travel against the arrow of time. Under most circumstances I would have to say that would be impossible but maybe I'm wrong.

Einstein has developed a dimension of time that contradicts the pathway direction and conception of the universe as a whole. Are wormholes possible vaguely I am leaning towards the idea that there is a lot in today's theoretical physics that aren't possible based on the evidence at hand? I believe that there is to some degree that time travel may be possible but the concepts for are it may incredible. Most scientist today rely on the gemology of Einstein's work based on his equation $E=Mc^2$ but it is incredibly true that space travel through wormholes that interfere with penetrating the space grid that all planets, stars and galaxy's lay at rest upon a spacecraft no matter what its velocity could not penetrate through it to reach a distance star or planet through a shortcut under the space grid. The grid holds in position enormous planets the brightest of radiated stars and the weight of galactic and nebula celestial grids. To form a wormhole to pass through this already rigid and depressed celestial grid floor planets and galaxy's lay upon seems impossible when reviewed under these circumstances. Space travel may only manipulate space by what I refer to in my earlier book "Albert Einstein's Universe" which explains "Instantaneous Velocity". It explains the foundation that the fastest a vessel can travel in space is measured by not only the default of a zero point gravity atmosphere but the ship's potential propulsion force.

Chapter Thirteen

"The Historic Flight"

Albert Einstein in 1904 came up with an example to show the effects of time that he called the "twin paradox". It begins with a pair of pretend twins, Al and Bert, both of whom are 10 years old in their highly futuristic universe Al's parents decide to send him to summer camp in the Alpha-3 star system, which is 25 light-years away (a light-year is the distance light travels in a year). Bert doesn't want to go and stays home on Earth so Al sets out on his own. Wanting him to get there as quickly as possible, his parents pay extra and send him at 99.999 percent the speed of light to Alpha 3.

The trip to the star and back again takes 50 years. What happens when Al returns his twin brother is now 60 years old, but Al is only 10 and a half now? How can this be? Al was away for 50 years but only aged by half a year. To Al he believes he has just discovered the fountain of youth?

The problem with the twin paradox theory is the object of time is the specific object to be reckoned with a long with the objector fact called velocity. It takes an alternative method of time to change the object of coordinative time and how can that be done. It is not traveling at the speed of light I assure you. It's more like having the ability to beat time itself to be on the real side using time as a coordinates and velocity as the means not the speed of light. To make TIME assertive units of a clock coordinate units to click time backwards one needs to travel faster than the time of the smallest measurement of time. One must travel faster than the measurement of the clock before it moves forwards to the next smallest Planck unit of time.

Each system of stars and planets or solar systems and galaxies all retain varieties of other planets and stars to act as whole system. No star or planet resides on its own. The time coordinates of the targeted star system has to be weighed, coordinated with its density a disk depression momentum that the systems reside as in a system as a whole. Its elliptical path depression and smallest Planck time intervals as a system must all be accounted for also. All this and more are not calculated by the primitive speed of light Einstein time travel theory. To travel into a planetary system's past historical history a ship would have to have the capacity to physically maneuver faster than the smallest degree of PLANCK TIME that exist somewhat before a space unit between the smallest calculable Planck time unit between the tiniest of time infinitesimal units that historical exist and that include whether or not such a space between time exist at all. This journey would also mean short cutting through already molded elliptical path space curves that lay between the stars and planets molded by their orbit time share continuums and is a type of historic flight only possible with futuristic advance space futuristic technology.

Einstein's idea about time slowing down sounds fine and dandy in theory, but the speed of light is actually a twentieth century primitive theoretical idea old in its mathematical configurations and how can we be sure for that matter he's even right at all? One way would be to hop in a rocket and travel near the speed of light. Yet everything we know about physics says we can't do it. According to Einstein's special theory of relativity, objects gain mass as they accelerate to greater and greater speeds. Now, to get an object to move faster, you need to give it some sort of extra push. An object that has more mass needs a bigger push than an object with less mass or even no mass as light retains. If an object reached the speed of light, it would have an infinite amount of mass and need an infinite amount of push, or acceleration, and even more to keep it moving. No rocket engine, no matter how powerful, could do this. In fact, according to relativity and as far as we know, nothing can exceed the speed of light.

Another problem with Einstein's special theory is that it takes a specific measure interaction of gravity for objects to gain mass and in space in the dark celestial realm even along the magnetic highway zero point energy minus $-g^2$ gravity is presented. Evidence in the magnetic highways discovered in 2011 sent by pictures from a satellite traveling past the four billion mile mark at the end of our solar system show particles traveling at undetermined speeds through the separation of the solar systems outer rotating disk grid and the next grid nearest to it. We should remember that relativity and all the experiments done on relativity were performed at the earliest time of the twentieth century and were experimented under the influence of gravity even the Michelson and Morley Experiment measured 'light' in a vacuum with no results of a massive object traveling through any type of laboratory vacuum tube at light speed.

Referring back to Planck time units' time shortens and does not travel backwards in Quantum Mechanics. The unit actually shrinks instead. But there must exist a point of interference between time intervals at any length where at what time might be called a 'universal Planck time units' existing as a point interval between measurements that a division between the arrow or flow of galactic-matter orbitration unit even at infinitesimal length measures a space by which time actually may unfortunately stop relative a coordinated short time space-time-velocity between universe momentum and time equals zero not historically but intermediate to the present that may act as the means for time travel which could allow the maneuver to travel backwards into the universes' past. To raise the possibility that a journey into the future might be possible is a legendary one considered an impossible possibility to even think of.

To travel into the past we coordinate a velocity to alter the older coordinates of time itself. To travel into the future we would have to re-coordinate a velocity even more so re-coordinating infinitesimal time unit intervals for the past journey that could help coordinate the ship's computer to travel into the future and it would take a much greater degree of thought.

In an atom the positive and negative components cancel each other out. The argument is;

"Orbits are the equilibrium of forces. One aspect of inequality is *massive differences*, but the forces are equal and opposite in Newton's mechanics". This is correct inside any particular atom, or if there were only one atom in the universe because the electromagnetic (electrostatic) range is infinite, *there is a very slight attraction between atoms.* So, gravity is simply the weak electrostatic attraction between atoms. It is simple to believe its all electrostatics!

The Attractive forces exceed the Repulsive forces because the Attractive Forces are coupled together and work in concert. The Repulsive forces are not coupled and do not work together. It's like two against one. But the repulsive also keep everything from collapsing. In Quanta Physics Theory are the celestial sphere size mass deep space coordinates it is not attraction that perceives as the force that the planets and stars abide by. In the curved space gravity well stalled celestials lay on proceeding against and laying on top the fabric of the space grid they impress positioning from it is the fabric grid that curves with the angular momentum cycle as a universal body and action that causes the curvature.

(See: The Foucault Pendulum Exhibit from Panthéon, Paris) As the smaller planets compared with huge galactic celestials it is a free fall action that is impressed holding a determinate position in the grid itself what Newton and Einstein perceived as attraction by matter-energy even though it abides astronomical distances between the celestial bodies and determines how close or far a distance the indifferences of celestial size masses reside it is the free fall action of the universe's cycle that does the work to impress the bodies within the fabric itself but it the repulsive forces that keep the bodies separated from collapse.

The *gravitational force* would be a small amount of force per atom. (10-41) this amount would vary depending upon the element mass of each individual atom determining what grade of mass it uncouples. If the moon was made of Lithium Element, it would have less gravity than if it was made of Uranium.

CELESTIAL FREE FALL INTO MOTION

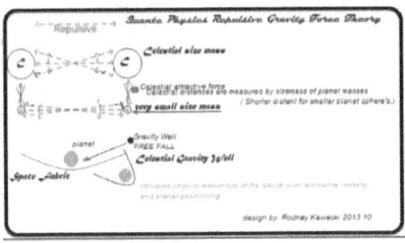

CELESTIAL UNIVERSE EXPANSION THEORY

Matter retains no physical velocity at rest in the air or in space unless a force is applied to its physical mass. The universe in Quanta Physics Theory as a whole is in its design an oval three dimensional bubble embedded in deep space bending, twisting, stretching and curving the fabric crest that positions it. Due to its physical mass size, chemistry and orbitration rotation cycling the celestial sphere or cosmic egg we refer to as the universe causes a complex 'weight' orientation maneuvering ability because it orbitrates in a momentum caused by a perpetual angularity because it is embedded in an erotic space field curvature. It twist and recedes inside its 'gravity-well it embeds in the invisible fabric of space itself.

It has been explained in classical physics over the decades to rotate into an angularism motion that it maneuvers in a circular arc that can be described in the same manner as the planets like our own do. Its angular shifting is due to a complex positioning from deeper depressed matter that scatter it that eclipse it into a physical arc momentum orbitration cycle due to the absence of surrounding matter that could if it were there create a repulsive resistance but since open space is ninety nine percent empty space allows the cosmic celestial sphere to reposition itself forming a specific dense trend of velocity. The Quanta Physics Expansion theory acts as an extension for "The Hubble Recession Theory" which explains an expanding universe deciphered through measurable astronomical distance lengths at the galactic scale. Quanta Physics explains universe recession due to the angularism of the universe's motion activity explained by orbitration momentum, perpetual motion and angular free fall motion that contradict them due to the upper and free fall lower cycling in the universe's rotation. Cosmic celestial matter in the universal sphere shifts its weight due to the momentum in the cycle physical creative motion that explains a 'wobble affect'. Beneath the 'cosmic celestial surface the fabric of space curves and stretches acting as a 'gravity well' our universe maneuvers in.

In our perpetual design we see a monishes orbital spin of our universe based on the majority of vast galactic bodies that lay across the universal plain fabric that due to the absence of gravity weightlessness allows these gigantic bodies to create universal motion. Because our universe explain here is a physical celestial sphere entity all the gigantic galactic bodies lay on a celestial crest gravity fabric ascended from the inside foreign chemistry of the celestial sphere that is our universe.

We look at the fabric of space in the same way as historians of the sixteenth century looked at earth believing at the time the earth was flat. The Celestial Universe Theory is described in the same manner. We look at a distance event horizon from a distance failing to observe what lies beyond it. As a universe singularity we invent the same type of event horizon not including what lies beyond it. Like the earth our universe is naturally round. The galaxy's that play the largest entities in it lay between them in gravity wells that decipher their own separation and distance at astronomical lengths. Due to similar matter poverties they recede at a distance based by the density of their energy that because similar repel but dependent to the polarity of the planetary star energy.

The galaxy's, Nebula and cluster galactic bodies are the atoms formed from the celestial cluster that lies inside the cosmic egg. Based on the perpetuity of the spheres motion inside the celestial sphere pressure and resistance of molecular deities duplicate and smash pushing against the fabric crest surface inside it that acts as a shell for the celestial body. Leaks crack the cosmic eggs crest empowering release of the chemistry inside the egg and becomes pushed out of it yet is still caught within the celestial's upper equilibrium surface realm where duplicating and vast molecular regenerating occurs in the pressure release process. Some leakage chemistry defer into gaseous nebula and other galactic mass properties forming other galactic star materials that summarize into deep gravity wells.. It is the presence of these leakages that lay upon the celestial universe surface fabric. It is the nature of the universe's spin that these galactic planetary models orbit and rotate.

Multiple universe theory resides abroad a vast region of *dark infinity*. The reflection of enabling bright 'quanta' shuts the door from darkness as the mystery of our universe reveals the existence of its creation way beyond the astronomical but is still embedded in a vast darkness above the celestial grid.

Quanta Physics Logo

"The Planets, Stars and Galaxies are atoms and the Universe is busting at its seams"

In the earliest days of the 16th century we the research of how and why objects celestial and abroad maintained a specific balance of activity which is now known as the theory of gravitation. We studied the broad celestial motion of gravity and the closer activity of gravity of our own planet's surface activity of it. In 1904 Albert Einstein acknowledged a discovery about space which is now founded as the fabric of space by which at the celestial scale planets and stars along with the larger galaxies lay abroad on top of in a specific design he called flat space.

Since the universe spins as do all the other common celestial bodies of the universe science was lead to believe that the universe acted as a disk plate that rotated And orbited in the same fashion as the smaller celestial bodies that resided within it. The universe spins dragging along with it all the other debris in a designated cycle where inside the larger galaxies solar systems and star systems retain the same pattern impressing them deep in this dark fabric we know as space.

In 1776 English declared the fashion that the world was flat also and sent ship's to discover whether or not one would fall off some far away edge of the world's flat surface plain. He didn't and the idea about our planet changed. The world is round the people claimed. When we view these same types of ideas about space we view space as being flatland in the same fashion as the English in the early times of the seventh century. Erwin Hubble discovered in the early twentieth century along with Einstein that it's possible that the universe acts like a balloon and is expanding at its wits end and proclaimed that the universe was indeed expanding. Likewise our own view of our own planet looks at its surface edge towards an horizon flat but we do know now that it's not.

Our universe may take on the same characteristics as this short story about the celestial plain and both are true. The earth is flat to the sight of the eye but is round in characteristic as the English believed centuries before. Studies in the field of science have leaded us to an era testing our ability to decipher whether or not a ship can travel faster than light. My own ideas on this are it can and my books are based on those facts that arise from that evidence. Indeed outer space is a flatland but reviewing the aspects as it show that that invisible harden yet real fabric the planets and galaxies lay and wobble on is too great to penetrate and believe a spacecraft can maneuver and drive a ship through wormhole to shortcut a distance to a nearby star.

With that in mind I view our universe amongst others that probably exist in the infinite space field in its vast coldness where atoms materialized from a deeply depressed invisible shape from the middle of a vast sphere orbital circular in forensic mass creates eruption of stressed compressive allusive liquid from inside it rips and pushes its way outside of it at its weakest fabric ends fusing at galactic intervals over time displaying a material spread of atomic cluster, nebula and galaxy matter from inside revealing the presence of a gigantic celestial type universe hidden amidst the dark shadows of space itself.

Gravity is a repulsive force at both the celestial planetary and galactic scale as well as the intermediate planetary surface scales abiding by the same rules of gravitation throughout the universe gaseous and alike. Alike or similar chemistry matter made of atoms at the celestial planetary scale repel because they are made of similar matter. It is this process that calculates the distance between all the celestial spheres galaxies, stars and planets and is measured relative to their size masses. The planetary gravity forces at planet surfaces makes the similar celestial energy from the planet itself push back outwards objects falling towards its surface natively called free falling masses and is calibrated by the planets repulsive energy mass the same resistance which creates the distances between the stars and planets at the celestial but pushes free falling objects and is called 'free falling resistance'. It is the similarity in an objects chemistry mass and solidity that the resistance occurs as a gravity force and which at a planet's atmosphere and pressure density that a planets dominion deciphers a specific free fall velocity as in Newton' gravity measurements. An object falling towards a planet's surface no matter what its size or mass will always fall towards the planet's surface at the same velocity 9.8 meters per second as the dominion repulsive gravity energy is what acts as the dominion force that governs all objects entering it atmosphere falling from inside it or intruding it from outer space. Escape velocity or 11,000 miles an hour acts as a velocity to measure the atmosphere compression sealed inside the earth's atmosphere between it and space and is depressed by the planets elliptical path which acts to seal the planets nature environment. Where gravity pushes against the incoming weight of the object free fall the planet's atmosphere determines and calibrates the objects chemistry mass and weight as it slowly comes to rest on the planet's surface.

The deity of the space fabric where planets lay curving and stretching in a constant toggle war amongst themselves this dark invisible fabric is neither torn, ripped or even broken through big the largest of celestial bodies that reside on it. We can assume that this invisible table that gigantic celestial galaxies reside upon is not the actual space as we believe of it but more so is the foundation of it. Abiding by this assumption we can safely say that open (empty) space is a layered atmospheric upper existing plain where what we decipher as zero point gravity exist and thus is empty.

As we have absorbed the force of gravity as a systematic résistance our planet chemistry retains and acts to slowing down intruding entities from outer space and or the action for object free fall using its capacity structure to control and protect itself thereby minimizing a surface objects speed within it we notice a special difference of a zero gravity interaction of material deities working in close range to our planet's surface space. I say this because I want to put a great evidence foundation for the next few paragraphs.

When we talk about the universe we easily assume our universe is a singularity of many universes' that exist throughout infinite space. When Einstein said that the equation for gravity equaled infinity I think this is what he meant. To reach further into the mystery about our universe and magnitude of its existence we observed and through our planets history have created assumptions that form misled empowered branches on our theoretical tree. It seems to have become theatrical instead. Wormholes and short-cuts, drilling through a specifically understood fabric gigantic planets, stars and even galaxies reside. The possibility of racing against specific unit ticks on a clock to say calculate a certain rhythm of the universe and dominates as a dimension the sole nature of its being to be a clock.

When we look at what we call the observable universe we see aspects of a flatland celestial inhabited magic plain in which dimensions our infinite in nature. What we don't see is our universe's event horizon is just that an observation or observable seeable line that over views a gigantic celestial sphere. Like earth in the sixteen hundreds the people believed the earth was flat and that if you maneuvered a ship through the ocean you would fall off its edge. When we look at our universe we observe the same observation. A celestial in natured fabric flatland believed to be infinite in nature. But what don't observe is the ideas that if the universe is a celestial sphere it exist in the same fashion as the planets and galaxies that exist on its surface. The universe is as infinite as the darkness that materializes an invisible edge this cosmic egg as historians' display it is in a premature existence. As a sphere taking in all the characteristics as of history's betray ling of it we travel upon an empty dark upper atmosphere of the largest celestial known. What lays inside this gigantic cosmic egg is well unknown to probably any nature of life in existence throughout this celestials boundaries.

We acknowledge its fabric surface as the foundation galactic and planetary alike celestials exist on stretching and curving around on its surface as the celestial universe's wobbles and twist these objects as it was set in motion by the vast weight of the matter in certain areas that cause an angular momentum. The millions and millions of galaxies, nebulas and cluster deities that exist on it has materialized a specific weight indifference that due to its vastness is unmeasurable by any real earthly means. The atoms in these galactic entities formed from leakages or releases of pressure in which our cosmic egg has clustered for us as planets, stars and galaxies that lay demented outside our cosmic eggs surface.

The separation from the cosmic celestial nature inside our universe has changed into infinitely sizeable atomic materialistic matter and different in chemical but act as a basis for future universe chemistry. The chemistry of what lays inside our Celestial Universe is willfully unknown and will possible never be addressed or venture through any type of magical wormhole adventure. The weight of the atomic nature of the cosmic egg's leakages of pressurization to protect itself from some type of cosmic disaster or explosion leads us to the probability of and nature of the big bang theory. With a fabric surface scenario by which universe sphere's push and pull in an angular motion subsides and detracts in an oval submission critically balances by multiple universe celestial existence forms an infinite display image of a magnificent and endless foundation of space and time.

We review the mobility about time we observe a single unit of planck time units make astronomical distances of the universe's rotation unimaginable. A single fraction could measure measurable in *comical* units that exist beyond planck theory.

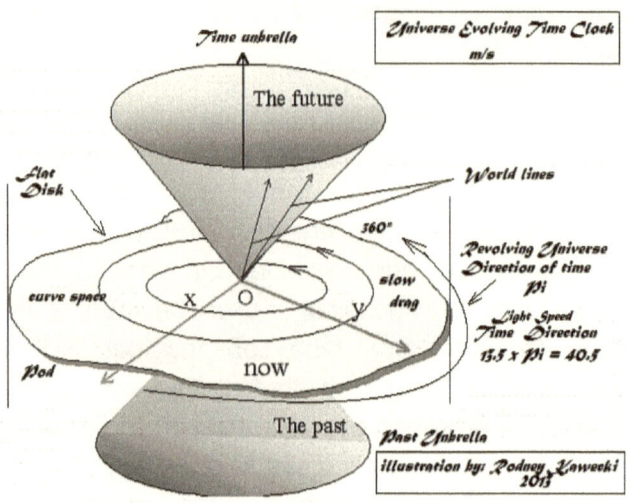

Time umbrella

The future

Universe Evolving Time Clock
m/s

Flat
Disk

World lines

360°

Revolving Universe
Direction of time
Pi

curve space

X   O   Y

slow
drag

Light Speed
Time Direction
13.5 x Pi = 40.5

Pod

now

The past

Past Umbrella

illustration by: Rodney Kawecki
2013

CHAPTER FOURTEEN

THE BLUE MARBLE EARTH

Designations for the planet Earth

Alternative names   TELUS or Terra, Gaia

Orbital characteristics

Aphelion   152,098,232 km

1.01671388 AU

Perihelion

147,098,290 km

0.98329134 AU

Semi-major axis   149,598,261 km

1.00000261 AU

Eccentricity   0.01671123

Orbital period   365.256363004 days

                  1.000017421 yrs

Average orbital speed   29.78 km/s

                  107,200 km/h

Mean anomaly   357.51716°

Inclination   7.155° to Sun's equator

 1.57869° to invariable plane

Longitude of ascending node

348.73936°

Argument of perihelion   114.20783°

Satellites   1 natural (the Moon),

8,300+ artificial (as of 1 March 2001)

Physical characteristics

Mean radius   6,371.0 km

Equatorial radius   6,378.1 km

Polar radius   6,356.8 km

Flattening   0.0033528

Circumference   40,075.017 km (equatorial)

40,007.86 km (meridional)

Surface area   510,072,000 km2.

148,940,000 km2 land (29.2 %)

361,132,000 km2 water (70.8 %)

Volume 1.08321×10^{12} km3

Mass 5.97219×10^{24} kg

Sun 3.0×10^{-6}

Mean density   5.515 g/cm3

Equatorial surface gravity   9.780327 m/s²

Surface 0.99732 g

Escape velocity   11.186 km/s

Sidereal rotation period   0.99726968 d

Rotation   23h 56m 4.100s

Equatorial rotation velocity   1,674.4 km/h (465.1 m/s)

.

Axial *tilt*   23°26'21".4119

Albedo   0.367 (geometric)

              0.306 (Bond)

Surface temp.

Kelvin   184 K 288 K 330 K

Celsius   −89.2 °C 15 °C 56.7 °C

Atmosphere

Surface pressure   101.325 kPa (MSL)

Composition   78.08% nitrogen (N_2)[3] (dry air)

                    20.95% oxygen (O_2)

                    0.93% argon

                    0.039% carbon dioxide

                    About 1% water vapor (varies with climate)

Human Being Density and Motion Mechanics

1062 kg/m3 = 1.062 g/cm3,

Average human body density 1408 kg/m3 = 1.408 g/cm3,

The average density of Earth is 5540 kg/m3 or 5.54 g/cm3.

Earth's surface, known as standard gravity is, by definition, 9.80665 m/s² (about 32.1740 ft/s2).

'It takes twice the volume in a mechanical mass to achieve an average maneuverable motion' that actually unifies in empty outer space and increases the velocity of the motion to twice the calculated speed.

The average density of Earth is   5540 kg/m 3

Average human body density is   - 1408 kg/m3

Element/Gravity/Weight   4 1 3 2 kg/m3

        Velocity   9.81 m/s

        Gravity   4.9 m/s (one-pound average)

        9.81 m/s or $mF = \frac{1}{2}$ mass + $\frac{1}{2}$ energy attraction.

The Final Frontier

'It only takes a single volume in a mechanical mass to achieve an average maneuverable motion' that actually unifies in empty outer space and increases the velocity of the motion to twice the calculated speed in *space*. *E = m°G (2.7c)*

Relativity Theory means e+v@c² = destruction but is propagated.

Hubble Space Expansion with 42 m/s every 3000 million miles.

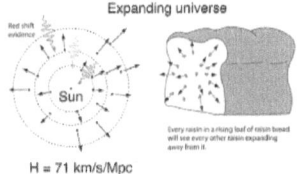

Reverse Vacuum Space Expansion equation   e+h @ g

2.7c = FTL

Chapter Fifteen

Reverse Vacuum Space Expansion

Reverse Vacuum Energy equation $e+h @ g 2.7c =$ FTL

Dark energy is the name given to the mysterious force that is accelerating the rate at which our universe is expanding.

We have known that our universe is expanding – that is, galaxies and clusters of galaxies are moving away from each other – since the observations of Edwin Hubble in the 1920s. It was thought that the rate of expansion would slow down over time as gravity gradually exerted a braking effect. However, about 20 years ago it was surprisingly discovered that not only is expansion not slowing down, it is actually speeding up. Some repulsive force is pulling the universe apart and this force was dubbed "dark energy".

Unlike mass, energy can produce an attractive or repulsive gravity depending on whether the energy pressure is positive or negative. The vacuum energy in theory has negative energy pressures, the problem with modern gravity theory is 'gravity' in space can be viewed as a physical action created by a physical thing called " The Expanding Void". An imagery of a round sphere embedded at the central mass of an expanding matter universe that is expanding into an idle space condition.

Known in The Quanta Physics Theory, space expansion is dubbed from an expanding universe theory illustrated by Erwin Hubble's research shown as an expanding universe entailed in three parts. 1) idle space. The condition of a reactive material or element that allows interaction of primeval forces. 2) The material universe. That part of the universe all made up of all matter existing throughout project universe and makes up of about 6 percent of everything. 3) The expanding void. A physical deity evolving mid-center the universe causing it to grow or expand.

Setting aside the factual basis of this expanding bubble as it has been called the void is reverse vacuum compression of the fifth element. Amongst the five primeval elements it is space in a concentrated form caused by the explosion known to us as the big bang event. Its cause is the effect of an empty vacuum affect suppressed by reverse engineering formed in the aftermath of creation.

This idea is equivalent to the cosmological constant used by Albert Einstein in his equations of general relativity and represents a constant energy density throughout space it is his notoriety of the "C' constant that was found to be invalid by the later Hubble Exchange. The idea of virtual particles per se in vacuum could also show the alignment of unknown territory in physics.

Repulsive Gravity

"In the theory of general relativity, we usually assume that the energy is greater than zero, at all times and everywhere in the universe," says Prof. Daniel Grumiller from the Institute for Theoretical Physics at the TU Wien (Vienna). This has a very important consequence for gravity: Energy is linked to mass via the formula $E=mc^2$. Negative energy would therefore also mean negative mass. Positive masses attract each other, but with a negative mass, gravity could suddenly become a repulsive force.

Even if the matter is somewhat more complicated than previously thought, energy cannot be obtained from nothing, even though it can become negative. The new research results now place tight bounds on negativity, thereby connecting it with quintessential properties of quantum mechanics.

If I went to space and had an empty box/container and went outside to fill that box with space, what would be inside the box?

1. Outer space, the void that exists between planets and stars, is not completely empty. Stars, planets, and moons keep their atmospheres by gravitational attraction. The density of atmospheric gas decreases with distance from the earth but there is no fine line

where earth's atmosphere ends and outer space begins. If you take a box as far as possible from the influence of heavenly bodies and nebulae, you can theoretically say you are in "outer space" but outer space will still contain a low density of atomic particles (predominantly a plasma of hydrogen and helium), neutrinos, and dusts, as well as electromagnetic radiation, magnetic fields, and cosmic rays.

2. Vacuum is space devoid of matter. It would be nice if we could say that a vacuum is space devoid of matter and energy but quantum field theory tells us we can't. There is always background energy called vacuum energy in a vacuum. The effects of vacuum energy have been experimentally observed in phenomena such as the Casimir effect and the Lamb shift, and vacuum energy is believed to influence the behavior of the universe on cosmological scales. Therefore, a vacuum is not empty space. It is the ground state of space with the energy of the ground state called zero-point energy or what's in Quanta Physics Theory is called "Idle Space". The latter does not mean zero energy per se but the minimum energy below which a thermodynamic system can never go. The source of vacuum energy are virtual particles or particle pairs that pop into existence and then disappear immediately such that they can't be directly observed.

For now, the only way physicists can analyze these virtual particles is through mathematical abstraction. Whether they have concrete physical existence is controversial. If you intend to fill a box with vacuum, you can only do this in theory. The vacuum "inside" the box (not the box itself) will have virtual particles and zero-point energy only but it also can contain a physical affect or reaction of unknown elements at the third quarter.

Reverse vacuum energy in theory has neither positive or negative energies. It is the empty box. The problem with modern gravity theory is 'gravity' in space is repulsive and can be viewed as a physical action created by a physical thing called " The Expanding Primeval Egg". An imagery of a round sphere embedded at the central of mass of an all matter universe that is expanding into an unknown yet idle condition of open empty space.

In the case of our universe this idle space action is an affect made by the big bang event itself. The false vacuum created by the big bang explosion expansion in an idle space condition that then evolves into model inflation, we discover a vacuum hole at the center of creation has been formed where the bang started. The big bang has done what it did with regenerating matter, it did also dig a hole for a reverse-vacuum in empty space. It created a reverse vacuum by absorbing idle space flexibilities into a concentrated space at the center of fusion. The condition that illustrates why the universe is physically expanding by a physically embedded and concentrated body.

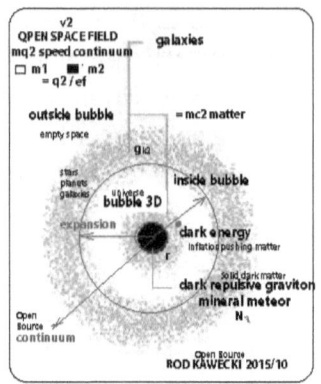

The Quanta Physics Theory

The figures are shown here:

93 billion light years (flat universe) astronomy dictionary
Albert Einstein GRT

Round-about three-dimensional mass – a multiple of pi.

279,000,000,000 light years (^{2}c) pi

3,000,000 light-years (minus a zero point for gravity) to maintain a substantial rate of velocity)

10,000 x 45 miles a sec (46.2 plus or minus 1.3 miles) = 450,000 miles a sec

Without using light frame measurement intervals"

(450,000 m/s) divided by L.S. = 372,000 m/s plus
78,000 m/s equals Speed without Limit

Equals: 44.7 miles per second every 3 million light-years

{450,000 m/s @ 99.999999 = 372,000 m/s} E=mg²
(2.7□)

Will Quanta Physics theory physically prove to show quantum field theory wrong?

A ship travels through idle space. It has the capacity to travel at the speed of light. Because of space expansion due to our expanding universe the ship travels 372,000 m/s. The Reverse Vacuum Effect caused by the flow of the universe's expansion allows zero mass (m°) receding vessels traveling through space be pushed away furthermore rather than be attracted by close planetary gravitation and so let's us travel through an idle space field at 99.999 percent capacity. This also includes faster than light velocities with no speed limitations because the reverse vacuum field gravity uses the equation $E=m°G^2$, instead of the latter $E=m°c^2$.

The reason is rather that gravitation in space is a reverse-vacuum natural dark matter push force, and rather hopelessly a physical opposite of the planetary attraction force.

With no exterior forces acting on our vessel except for the universes space expansion volume force itself we are able to travel the full acceleration value of 99.999 percent velocity having nothing there to slow us down.

Venturing into relativities infinite mass theories, we resume the value of the uncritical assumption of 'nothing can travel faster than light" due to the attraction field forces to the indifference principle with the expansion value that the ship that travels light speed due to the conditions of reverse vacuum response now travels twice that velocity based on the standard nature. What attraction theory does to slow things down the reverse vacuum response accelerates thing up faster.

As the capacity of acceleration rises due to the advancement of technologies our universe will continue to expand getting larger and larger over time. The idea that we will ever be caught up with *god's creation* is overwhelming. Today big "G' allows us to travel 450,000 miles in a single second of time but in the future this number will change. The idea that mankind reaches all the criteria in its engineering propulsion engines that will exceed this number relies on the idea that a reverse push vacuum expansion force allows an expansion of primeval accelerations can be measured with the foundation that the universe space field comes with its own advance acceleration condition that when or if reached will take more propulsive engineering advancement technology to oversee the greater velocity.